Elon Musk

The Business & Life Lessons of a Modern Day Renaissance Man

(Business Principles From the World's Most Powerful Entrepreneur)

David Hill

Published By **Regina Loviusher**

David Hill

All Rights Reserved

Elon Musk: The Business & Life Lessons of a Modern Day Renaissance Man (Business Principles From the World's Most Powerful Entrepreneur)

ISBN 978-1-77485-836-3

No part of this guidebook shall be reproduced in any form without permission in writing from the publisher except in the case of brief quotations embodied in critical articles or reviews.

Legal & Disclaimer

The information contained in this ebook is not designed to replace or take the place of any form of medicine or professional medical advice. The information in this ebook has been provided for educational & entertainment purposes only.

The information contained in this book has been compiled from sources deemed reliable, and it is accurate to the best of the Author's knowledge; however, the Author cannot guarantee its accuracy and validity and cannot be held liable for any errors or omissions. Changes are periodically made to this book. You must consult your doctor or get professional medical advice before using any of the suggested remedies, techniques, or information in this book.

Upon using the information contained in this book, you agree to hold harmless the Author from and against any damages, costs, and expenses, including any legal fees potentially resulting from the application of any of the information provided by this guide. This disclaimer applies to any damages or injury caused by the use and application, whether directly or indirectly, of any advice or information presented, whether for breach of contract, tort, negligence, personal injury, criminal intent, or under any other cause of action.

You agree to accept all risks of using the information presented inside this book. You need to consult a professional medical practitioner in order to ensure you are both able and healthy enough to participate in this program.

TABLE OF CONTENTS

Chapter 1: Eldest Child Of An Engineer Or A Model 1

Chapter 2: Droopout Zips Insert Into The Industry Wizardry 13

Chapter 3: Competition For Person To Person Trading.. 26

Chapter 4: Getting Private Space 41

Chapter 5: The World By Being Sexually Attractive 48

Chapter 6: Thoughts Coming From A Festival... 59

Chapter 7: Flirting To The Second 64

Chapter 8: Little Known Facts 125

Chapter 9: Entry Into The U.S................ 138

Chapter 10: Developments And Innovations 176

Conclusion 183

Chapter 1: Eldest Child Of An Engineer Or A

Model

Elon Musk was the eldest son born to Errol as well as Maye Musk in Pretoria, Transvaal, South Africa on June 28th 1971. Elon Musk has a brother named Kimbal Musk, who would play a part for his older brother's successes and a younger sister, Tosca Musk. Errol Musk was an engineer, pilot as well as a sailor, as a sailor, pilot and engineer in South Africa, and Maye Musk was a popular model and a finalist at the Miss South Africa Beauty Competitions, and a dietitian. Errol Musk was the son of an British mother as well as a South African father. Maye Musk came over to South Africa with her family from Canada in the year 1950. She has Canadian citizenship.

From a young time, Elon Musk displayed intellectual talent. While most children would wonder over the world of nature and be awestruck by things they would never reach or touch, Musk had a different perception.

While others claimed that the moon was one million miles away. Musk declared it closer to 200 miles. It left other kids in disbelief at Musk's sense of wonder but it was a sign of one of the most impressive qualities of Musk his ability to accept absolutely anything seriously.

The other kids were definitely less social as Elon Musk. While they didn't have the same enthusiasm for academics or technological skills They were given the opportunity to have a great time. Instead of creating friction between siblings, it was beneficial for Elon. Through a positive relation with the siblings Elon was able to adapt his insufficient social abilities.

Musk's school was not his most favored subject of interest, neither. Instead of learning alongside teachers or tutors Musk was a child who Musk was a voracious reader. Musk claims that after the library ran out of books in the library suitable for his level of reading and he started to read the Encyclopedia of the World.

When he was ten years old when he was ten, he started to take an fascination with computer programming thanks to his Commodore VIC-20. He says that he did not have ambitious goals as a young person and that his progress into technology was driven by a love of video games. It was his love of games the reason he achieved these first-rate achievements. According to Kimbal at the time, Elon was tested by IBM and IBM, who found out that Elon was one of the greatest abilities in computer programming they'd ever witnessed.

At the age of 12years old, Elon was able to teach himself to program. He was attempting to attend classes on computers but it wasn't because the class was not ahead of his peers; it was because he was far ahead of the instructors. It was not worth taking the classes. Instead he resorted to what was always worked for him: books. This helped him program his first hit as an entrepreneurial entrepreneur: Blastar. It was a game that he designed with the help of the information could be picked from the library. It was sold at

$500, which is around $1300 in value, accounting for inflation in 2019.

In the same year Kimbal and Elon were planning to build an arcade close to their school. After getting a lease accepted for the space, and a firm willing to supply the equipment the only thing they needed was a city permit however, they needed an adult to apply for the permit. The boys had already shared their ideas with neither nor their respective parents. After they approached their parents about their business plan they were not able to convince them of the idea.

Despite his intellect or perhaps due to his intelligence, Elon was bullied. Due to his birthday and ability to think critically, Elon was allowed to begin school one year earlier. This meant Elon was the shortest and the least experienced of his classmates. Instead of being acknowledged for his accomplishments and accomplishments, he was ridiculed for his academic progress due to the fact that he was less physically mature than his peers who were not as advanced to him in the intellectual sense.

Instead of being a slave to other children, Musk invested his time to study and read to enjoy himself. Comic books were among his most loved books to read and was hilarious considering how closely his personal story is a direct reference to a story straight from Marvel. Another work that was incredibly important to Musk came from the works that was done by Isaac Asimov, one of the Big Three of Sci-Fi alongside Arthur C. Clarke and Robert A. Heinlein. Musk spent his childhood reading Isaac Asimov's Foundation series, and says the work he absorbed taught him the following lesson: "You should try to follow the sequence of actions that will extend the lifespan of civilization, reduce the chance of a dark age and cut down the length of a dark age , if there's one."

His mother Maye says that he is the typical person who is a nerd. She remembers that during the early years of his life, the family could count on him to provide answers to their general inquiries and to be knowledgeable about the world surrounding them. Elon is a remarkable memory. Elon's habit of reviewing dictionary widened the scope of his knowledge. Tosca was his sister. She was beginning to feel a little off by his intelligence and when a problem came up to her family and friends, she'd reply at her mom, "Why don't you go to him and ask him a genius question?"

In school, it wasn't as petty or jolly. Elon was not just an object of ridicule for his eccentricities, but also for his physical insufficiency. Being a typical genius, his growth in intellect was a single-minded journey through his life. Instead of going out, engaging in games, or playing with drugs, Musk's youth was focused on gaining a better comprehension of the concepts Musk could comprehend.

However, this didn't provide immediate protection. In his school days, Musk was thrown down the stairs by an gang of bullies. They dominated him with numbers, and then utilized their advantage to throw him down the steps, onto the concrete floor, where they began to surround him. Once the cowards had him weak as well as outnumbered smash his head into the concrete. This went on until Musk stopped breathing. Then the patient was taken to a hospital. Then, he would have undergo plastic surgery on his nose to repair the injuries.

One could only imagine the impact this has had on Musk. His social issues are common among those who are intelligent however, instead of being inspired to explore his interests as a young person Musk was pushed away.

Musk However, he has done more than bounce back from the battle and, to this day, he is not having any issues creating connections or manifesting his ideas. At the end of the day, his exceptional qualities that helped him stand out , are exactly the qualities that make him an incredibly powerful humans to have ever been born.

Instead of accepting the gangs of bullies that would "literally chase him down", Musk used his ability to study in order to gain an advantage. Musk cultivated his passion for Karate, Judo as well as wrestling. "I began to distribute it at the speed they'd offer the opportunity," he explained.

Musk explains his experience. Musk states, "It taught me a lesson: when you're fighting a bully whom you can't please, you hit the bully's nose. Bullies look for targets who will not resist. If you are an easy target and kick the bully's nose and he's going smash the hell out of you, but not likely to do it once more."

It was 1979 when Errol Musk and Maye Musk divorced in 1979. It was a painful time for the couple and for any family who has had to go through it. Errol's kids have described Errol in unflattering manner. Although he opted to stay with his father for two years after divorce Musk believes that the move was an error. Musk refers to as the South African Engineer as "a awful individual".

Elon was able to move into the home of his father because Elon was jealous of Errol and was upset that his mother was the one to get three children as part of the divorce. Elon wanted to ensure that his dad didn't be lonely and sad after the divorce. In spite of all the kindness, Elon, unfortunately, ended up learning the exact lessons he was taught by his bullies.

When he talks about the father of his "He was such a horrible human being, you do not know. My dad has planned out a meticulously thought-out strategy of evil. He'll plan an evil plan. It is impossible to know the extent of his evil. Every crime you could imagine that he has committed. Nearly every sin you can think of that he's committed. It's so horrible you'll be shocked. As I've seen it there's nothing to do. Nothing, nothing. I would like to. I tried every trick. I tried threats, rewards emotional arguments, intellectual arguments all to change my father's behavior to better. And I was unable to change him... There was no way. There's no way to change it, it only went on to get worse."

It was never about work with Elon. Elon was a hard worker in all the activities he was involved in. If he was forced the option of fighting, he took part or demonstrate determination, he did. His attitude did not lead him to aid his father get to a better life however, Elon's strength and determination helped keep Elon from being shattered because of the lack of respect his father showed his son.

Through his childhood, Elon looked to America particularly Silicon Valley, as an almost mythical destination. This was the place he envisioned as the place he envisioned as Mount Olympus, which he often imagined through the experiences of being raised in a country which didn't have the same opportunities to develop technology. In a way that is heartwarming, Elon has always looked at America for the job of possibility which is where the most amazing things occur. This belief fueled him during his studies, and Elon discovered more and interesting research within his areas of research.

In the opinion of Elon the fields that were worth his time included the internet and electric vehicles, space travel as well as artificial intelligence and genetic modification. Born into a family where he had a passion with video games Elon found himself most interested in the electric vehicle, the internet, along with space-travel. The most innovative developments in computer and physical engineering were being developed in California.

After graduation from high school shortly before turning 18 years old, Elon Musk made the decision to embark on an important life-changing decision. Elon impressed his mother with his show of self-confidence. Contrary to what his parents had advised Elon was granted an international passport and took a flight to Canada. He was aware that it was easy to gain entry into America from Canada and that's why Elon chose to go there. The reason he could relocate to Canada without difficulty is because He was able to obtain an Canadian passport by virtue of her Canadian citizenship. When he arrived, Elon called his mother and asked her to tell her what to do.

Elon was able to take his bus ride that ran from Montreal to Vancouver and back, which gave him to take in the country, although through the windows of buses and highways. Elon was energized by the shift in continent and was making progress to achieving his goals. When Errol called his father, Errol's mood was not reflective of his accomplishment.

Instead of celebrating the journey of his son, Errol called Elon an fool and stated that Elon was likely to return home to South Africa within three months. Errol continued to claim that Elon could never do it , and that he would never be able to achieve anything.

After working at odd jobs across all over the nation, Elon ended up getting accepted to Queen's University and stayed in Canada. This enabled him to stay out of compulsory military service with the South African military, a government that neither he, nor his siblings believed in. After six months, Elon left for Canada, Maye followed him there, bringing with her both of her siblings along with her.

CHAPTER 2: DROOPOUT ZIPS INSERT INTO THE

INDUSTRY WIZARDRY

Following two academic years studying at Queens University, Musk left Queens to pursue a higher studies at University of Pennsylvania. Through this he fulfilled his goal of making the move to America. He earned an economics bachelor's degree and later earned a second bachelor's degree in Physics.

When he recalls his time in college, Musk admits that he was not a regular student. Instead, he would stay in his bedroom, reading the compulsory texts and passing the tests. The most important benefit that he thinks school offered him was the opportunity to get to know girls of his age. In Queen's University that he met his first wife, Justine Wilson.

Musk's talents didn't perform as well in the academic world However. His talent was not in the academic abilities of professors or the fervent pursuit of an area that was that was locked in a lab. He had instead shown skills that stood out and in the marketplace. His remarkable ability in areas that few are skilled in and his knack for analyzing the trends of the market and the needs of consumers technology are skills that develop better when practiced than formal training.

After completing his degree within Pennsylvania, Musk got a award to go to Stanford at California. It would put him in close proximity to the rapidly growing industries that have inspired Musk since his childhood. In the absence of going to classes at the college, Musk called the chairman of the department and informed him that he would leave the university. Musk explained to his boss that he was taking the decision to start a business which was an online business and was certain to fail, which is why he affirmed his right to come back to school in the event that his plans did not come to fruition. Although he used it to provide security measure, Musk didn't trust that groundbreaking research was taking place in the college. He was in Stanford over two weeks.

In the beginning, Kimbal lived in Canada and Elon phoned him and advised Kimbal to come to California to begin the company on the internet together. Elon was just 24 at the time, and had very few connections within the state, as well as approximately $2000 in the bank. The conditions of living they endured were so minimal that they had to rely on showering in the gym. "[There were] a couple of instances when couches were available in the day and fraternity beds in the evening," Elon remarked when describing the way they lived during the period.

Through the money the brothers gathered via angel investors and angel investors, they were able to establish their own business. As they did what they could, Elon would write the program they were creating while Kimbal took care of the marketing. One of them created the product, and the other would sell it. The program they were creating was aimed at digitalizing local businesses, in essence making it possible that you could search for businesses on the internet.

It was a different age back in 1995, when people used maps for driving. It was a different time when the Yellow Pages were a hegemonic system for listing cities and their locations. people were envious of the brothers' attempts to modernize the system. One person walked to an office, and dropped the copy the Yellow Pages on the desk. He asked what they believed they could ever be able to compete with to it. The brothers weren't sure what to say because, they didn't just plan to be competitive with the Yellow Pages, they planned to take over the Yellow Pages.

Elon could still recall the basics of programming from the Blastar days, and so Elon wrote the programs. One of them allowed users to view maps on the internet. He also wrote the code for a program to alter online content and functions in a manner similar to the modern-day blogging. It was a groundbreaking tool in the late nineties, before instant messaging was popular.

Global Link, the business, Global Link, was run by a staff located in Palo Alto, California. The place was also where Musk brothers were sleeping. The office was situated above a small Internet Service Provider and they could get an affordable internet connection through drilling a hole into the floor. They had only four staff members. The one who was there was a close friend of their mothers, and the three others were salespeople were hired via advertising. Because they were already struggling with the burden of student debt, Elon did not want to invest in a house which is why when he discovered that he could purchase an office at a lower cost then he decided to go with it.

Through this arrangement, they introduced the product to a variety of companies in Silicon Valley. The majority of businesses were not aware of the internet in 1995. Because many Silicon Valley businesses weren't savvy to the internet in the year 1995 The company was not able to gain customers in the initial stages. There was plenty of work to be done, and some companies reacted to the notion of being featured online as "the most stupid idea they've ever heard of".

It was a tough battle with Elon along with Kimbal. They had to make it in a world where few people used the internet and there was virtually no possibility of making money from it. There were times when it was fair to doubt their decision-making process however the brothers remained in their convictions. They didn't let failure go unnoticed.

Then, they began to attract a fair amount of customers and the business began to level out. Global Link saw minor success and, given their minimal budget, it meant much. Because of their low expenses and their ability to operate with a small staff Global Link was able to create an income stream that was positive. This meant that the tiny profits they earned were enough to keep the small-scale business on the right track.

In the year 1996, the company was able to secure its first significant victory in the form of a three-million dollar investment from Mohr Davidow Ventures. It was at this time the company was renamed Global Link. Global

Link was then renamed Zip2. The program's fundamentals until that point were based on data from an Palo Alto business database and an Navteq database. Davidow changed the way he approached this by focusing on the market in general by selling back-end software to newspapers instead of direct-to-business sales. This means that Zip2 would be selling the tools required to build directories online to newspapers, instead of taking over them. The directories would then be able to create the maps displayed on websites and newspapers could create their own pages.

In the year 1998 Zip2 came up with the slogan, "We Power the Press." Zip2 had signed deals together with The Hearst Corporation, The New York Times, The Chicago Tribune and about another 160 newspapers. Musk states that 20 of the newspapers were able to create complete city guides, which brought these into the new digital world. In the next two years, it would be the norm for business however, in 1998 newspapers were just starting to develop their online magazines.

This huge success was not what Elon wanted however. When Mohr Davidow was more involved in Zip2 as they took over responsibility off of Musk. This enabled him to lead more comfortably however, it also meant that he was no longer the sole owner of the code, or Zip2 as a company. Mohr Davidow hired more experienced engineers to enhance the initial program. They reduced the bulk of Musk's work, and cut down on the bulky code. This was a problem for Musk since he had developed the company and program by himself.

However, the recently successful Musk wasn't as effective than Mohr Davidow. Vice-president of engineering Jim Ambras, knew to provide Musk some time to work around his schedule. When Musk said something would be completed in two hours, which means it will take a whole day. If Musk said things would take a day, it could take one week. His abilities to manage time weren't fully developed at the time, and he didn't possess a complete understanding of his abilities. Maybe, Musk didn't enjoy the flexibility he experienced when working alongside five other people However, the experience of the larger organization ensured that he was able to keep Zip2 in good condition at a national level.

This enabled them to enjoy amazing success. They were the pioneers of the electronic industry, creating the core functions for the web. At a time when many believed that the internet was an unpopular trend, Zip2 proved that there existed a legitimate function for this World Wide Web, and most importantly, they proved that there was a way to earn from it.

On February 29, 1999, Zip2 was sold at a price of $307million in cash and cash Compaq Computers. The purpose of the sale was that Compaq could upgrade the search engines they use, AltaVista. By selling his shares that included within Zip2, Elon Musk made $22 million during the deal. Kimbal Musk made $15 million. This was their first huge achievement.

Then, almost immediately following the purchase, Elon bought a McLaren F1 which was one of only 64 vehicles in the world at the time. His family was a fan of racing and he was eager to quit his job and travel the world. The car was brought to his home, Musk had invited CNN to join him in welcoming the car into his home.

Justine stated in her interview she didn't want the money to alter them and that they didn't want their children to become spoilt. Musk expressed her views by saying: "I'd say the real benefit is the satisfaction that comes from having created the company that I later sold... however it sure is fun to drive." A few years prior Musk was in a bed on the floor of his office in 1999. By then the year he had the world's most expensive car.

The interview shows a distinct Musk is revealed in the interview than the one well-known in the last decade of the 2010s. He is an aspiring young man overjoyed by the success of his experiments, but not frightened by the difficulties which he was confronted with. The signs of stress show in his face and posture however, he displays an inner energy that reveals the desire to interact in the global community on a large scale.

The content of CNN has been changing over the years. Through the years they've attempted to portray Elon as a reckless billionaire, who always puts his businesses and capital. They have tried to depict him as a dangerous eccentric who the general public cannot connect with. In 1999, this was not the case. The interview is more of a snapshot of life. It is an appreciation of the incredible achievement by an South African who moved to the US with nothing and then embraced the American dream.

In his own words rather than buying an island within the Caribbean, Musk was more focused on having an impact upon the globe. The car was beautiful however, it was really nice. Musk was determined to see the world improve, and that's the reason why he created Global Link. The result included millions of dollars in revenue and an important shift in newspaper technology, but it wasn't enough when compared to Musk's real passions. Musk wasn't a person who wanted to be in the shadows. He wanted to help the public with methods that were legitimate and, more importantly, was legally valid.

This is the reason why the CNN interview was created. It's a flimsy story about an eccentric millionaire receiving a brand new gadget and an opportunity to allow Musk to showcase his latest initiative, X.com. This company focuses upon one of the most transferable types of data available around the globe and modernizes it into a simple service, rather than having us depend on the physical interface to access numbers. It will transform the way in which the world operates and save the time of business and individuals.

This data is considered to be currency. After experiencing the success that made him a millionaire prior to reaching the age of 30, Elon Musk went on to begin a new venture to modernize another system. After granting newspapers the power to build websites for listings of physical businesses Musk would like to give consumers the capability to make transactions on the computer. So, we've got the birth of X.com.

Chapter 3: Competition For Person To Person

Trading

Musk was not looking to just change the way the internet works but to change it. The goal of X.com was to establish the first bank in the world which could compete with fiat currency. Money is simply entries in a database therefore, by transferring that data to web pages, it offered customers greater freedom in their money.

To help him establish this company included Greg Kouri, an individual who had a deep respect for Musk and had a great relationship with Musk. Harris Fricker and several engineers were also invited to join the business. A large portion of Musk's fortune was invested in the development of the business. Musk depended on Fricker to help him gain experience in the banking sector.

It didn't go well for the company, but. The two of them Musk as well as Fricker had their own ideas about how to proceed with X.com.

They tried to compromise and maintain the company's unity however, Fricker put an ultimatum the face of Musk. Either Musk will allow Fricker to run the business, or Fricker was going to leave to set up the company of his own based around the same concept. Musk did not agree with the requirements, and Fricker quit, taking a lot of the top engineers along with Fricker.

The company is now demanding even greater amounts of the wealth he has through the form of investments Musk had been carrying the bulk of the responsibility through X.com. Musk was required to hire the entire staff from scratch following his departure from Fricker. The team was required to work hard in order to bring the vision into existence during the next three months. Musk himself would often spend 40 hours straight in order to achieve this dream.

The night prior to Thanksgiving 1999 X.com began its journey as an fully operational and licensed bank that could transfer money online, and also centralize via a super financial services. The company was in relationship with Barclays the multinational based in London. investment bank. The focus over time was to make money transferable via simple emails, at a minimum publically. The business's success was based upon the services for financial transactions X.com could offer in the early days. By the end of the year there were 200,000 customers.

The reason there was this demand in the market was due to the fact that while debit and credit cards were increasingly used throughout the nation but checks and fiat currencies continued to dominate transactions. When it came to transactions made using credit cards, the customer must register as a merchant on the application they used, or else the only method to get payment in the event that you failed to make enough profits to qualify as a merchant was via the postal mailing system. This method wasn't suitable for many reasons. One of them is the fact that it was not efficient. It could take one week or more to complete a transaction and there were a myriad of issues that could happen in the handling of the currency or the check.

Innovating in this market with a payment system that was as simple to transfer via mail, X.com was a welcome alternative for the smart consumer. Particularly for those who sell online. But, X.com could have invented online banking. However, it was not the first company to be involved in online transfer of money.

They were in direct competition with Confinity, a company Confinity with a name that is derived from two words "confidence" in addition to "infinity". Confinity was a company that targeted customers of their Palm Pilot, a popular form of PDA which is also known as a Personal Digital Assistant. These devices anticipated the capabilities of smartphones. They were tiny handheld devices that provided connectivity to the internet and mobility. Confinity began to take advantage of the market and attempted to make it simpler to move money among Palm Pilot users, creating an application to the device, PayPal. The focus was on this for

Confinity, however they also provided an alternative method of paying outside of the app to everyone who uses the internet as well, referred to with the title PayPal. This was not the primary goal of Confinity however it was a no-cost system to transfer funds.

Both companies offered incentives to users to sign up to their services. X.com offered users 20 dollars for signing up to their bank, and X.com would give them $10 for each referral they made for other people to sign up with X.com. They also offered free personal-to-person financial transfer. Confinity also had a similar plan by offering a referral fee of $10 and only $10 to join their service.

X.com offered a greater capability than Confinity due to the fact that X.com is a banking institution and Confinity was basically a service that was person-to-person. It didn't affect their success in the marketplace however, as X.com offered a stronger product, Confinity had the better advertised one. A large portion of their marketing was focused on the massive website eBay.

One of the biggest websites of its time and indeed, eBay had over 10 million users in 2000. While eBay was around for five years by that moment, it was its initial period of rapid expansion. In 1999 there were 2 million customers, meaning they were expanding rapidly. In 2000, they had hosted an average of three million auctions every day.

In 1999, because of the restrictions on merchant accounts on the site eBay, users of eBay were restricted to coordinating transactions through check and money orders. This made transactions slow, even though the number of transactions was being conducted every day. Confinity discovered this market and capitalized on the opportunity by creating a logo in HTML for their service of person-to-person. The result was that they were able to expand because of the high demand for digitalized services for money.

X.com was aware of this and moved into the market following them with its own brand and services on the website. The two companies began to battle for dominance in processing payments on eBay. Confinity began using an automated system that would inform specific sellers on eBay prior to bidding, telling them on the auction that they were buying items to benefit charity. should they win the auction the seller had to join PayPal to receive the money. The company spent over $100,000 every day appealing to customers. They earned the biggest brand name on eBay as well as the highest position in the market in terms of traffic. However, it also meant they were burning money.

X.com did not grow at the same rate but it stuck to Musk's concept. Maintain a low overhead and, if you are able to keep it below the revenue. This ethos helped him go from working in a cubicle to becoming a millionaire and maintained X.com as a business with funds to spare.

All of this was happening without eBay being aware. Realizing the potential for the platform they purchased Billpoint at the end of May in 1999 to create their own personal-to-person service. In the same month, eBay would develop their own service that would compete with other platforms. The service was launched in May.

Since X.com was profitable but less popular service , and Confinity was popular but not sustainable The two businesses were intimidated by rivals. They decided to combine to form one company in March of 2000, but keeping its name X.com.

Bill Harris, a previously successful CEO of technology was appointed to become Chief Executive Officer of X.com. Peter Thiel, the founder of Confinity became the the Senior Vice-President of Finance and Max Levchin retained his position as the Chief Tech Officer. Musk was the biggest shareholder of the company assumed the chairmanship of X.com.

This shift in power structure did not benefit Thiel or Musk. Harris's experience did not translate into the fast-paced and competitive atmosphere that the two companies were looking to enter prior to the merger. Instead Harris had the company in a state of bureaucratic chaos and hampered the expansion of the company.

After the merger, Harris directed the strategy of the business to reward PayPal customers who sign to the X.com savings account. The idea was tested over a period of two months. However, it failed to have any results So Harris came up with a different concept. He was planning to start charging PayPal users PayPal on a monthly basis for this service.

Thiel was against this approach from the start. He didn't believe in charging usersfor services, since it was against the premise of the company that it was intended to serve the customer He also believed that the method of promoting banks wouldn't be effective. Harris took the blame for Harris for X.com's course to become an unprofitable company. The last straw was when Harris made a political donation of $25,000 using company funds , without consulting with anyone other. After contributing $100 million to the business, Thiel stepped down as Executive Vice President.

Musk recognized the instability in the company and the detrimental effects it had on the executive. Instead of letting the company continue to run its course, Musk held an emergency board meeting, where he was given the authority to eliminate Harris out of the business. The board granted Musk the authority to remove Harris as well Musk took away Harris from the business, and later took over as the CEO of X.com.

In a show of faith and goodwill, Musk handed the chairman post to Thiel and he reformed the company's mindset. Harris had caused the company to believe that it would fail however Musk was capable of saving it and restore faith to the company's staff.

The contest grew as eBay collaborated in partnership with Wells Fargo, giving the bank 35 percent of their service, Billpoint. The partnership enabled eBay to offer customer support as well as back-end processing of payments. The time was when eBay also rolled out the Billpoint logo across its site and made it the most prominent element on all pages. Additionally, they made it more difficult to offer PayPal as payment options and restricted to the sizes of their logos allowed on their website by shrinking the

PayPal logo to only one-fourth of the size it was before. They also hosted Free Listing Days for whoever who used their Billpoint service, and further promoted the service on their website.

In the meantime, fraud ate into X.com's profits. By stealing credit cards, criminals who were computer-literate, mainly Nigerian and Russian criminals were able to direct take cash out of accounts via PayPal. The responsibility fell to those who sold the cash as consumers were able to collect the cost of damage from X.com. A group of criminals caused them to lose more than $5 million dollars in total. In June the year 2000 X.com had burned more than 10 million dollars each month.

Additionally the fact that they achieved success were unable to manage. The site that Levchin developed wasn't designed to handle traffic greater than two million users. They needed to design an entirely new platform that would permit more users to take advantage of the website. In the course of this revamp, Musk made the decision to change the platform the site was constructed on from the Oracle platform Confinity employed to the platform X.com utilized, WindowsNT. This resulted in a split between him and the majority of the founding Confinity engineers and employees, including Levchin.

The same was true of the debate over branding. Musk would like the entire business to use the name X.com which would eliminate the PayPal name. A significant portion of the workforce was dissatisfied with the decision, and tensions increased. When the rebranding process began, Musk left on a trip to the Olympics.

He was absent for two weeks on what was supposed to be a honeymoon, but turned into an opportunity to raise funds. His wedding was scheduled at the beginning of this year however the couple couldn't find the time to spend time with their relationship because of pressures from the business. While on Musk's travels, David Sachs, an executive, was not willing to be a part of the company's rebranding. Sachs gathered a group executives and threatened the board with threats to quit the company if Musk was removed from his post as CEO. While he was away the board members voted him out. A few Musk fans tried to reach him but the time was not right to be any impact.

When Musk came back, the former president attempted to convince Board members of the company to reconsider their position but ultimately decided it was not worth the effort. Musk states his position by saying: "Basically, I could fight it very hard however rather than fighting the issue at this crucial moment better to give in."

At the conclusion of the year, Musk was invested in the growth of the company. While he wasn't the chief executive officer but he was the sole investor of X.com. He was concerned about the future of the company and kept talking to Max Levchin and Peter Thiel. Thiel succeeded Musk as the CEO for the business. They soothed Musk's concerns and took charge of the venture.

At the time, Musk didn't need to be involved with the business. At the end of June, 2001

X.com switched its names to PayPal. In 2002 when they were listed on the stock exchange, they made $63 million. About a fourth of transactions that were conducted through eBay were handled by PayPal and more than 70% of auctions used it as a means of payment. PayPal has become the most popular payment method in the world.

Musk attributes a large part of this success to possibility that PayPal could offer an entirely separate credit card, one that let people access online funds from ATMs or outside of town. This function made it possible to transferable money that was available both offline and online and all via an account that functioned exactly as the cards consumers already had.

However, it was difficult to endure for this long as a rival to the users on platforms against it self, but PayPal was able to do it. On July 2, 2002, eBay bought PayPal in exchange for $1.5 billion, resulting in many millionaires who would later define what the next generation of internet. Musk was the biggest recipient of the profits, taking home $220 million and $180 million after tax. At this time in the course of his career, Musk aged 30.

Unsatisfied with his work, Musk immediately set onto new avenues. Even with all that had transpired that day, his new ventures were able to receive funding from Thiel. To avoid causing jealousy or animosity between the people who he worked with and him, Musk preferred to keep his work professional and open to working through the relationships Musk had built.

CHAPTER 4: GETTING PRIVATE SPACE

When Musk was looking through NASA and NASA, he could not discover what he was looking for. His most loved book as a youngster was The Hitchhiker's Guide to the Galaxy A hilarious tale that depicts humans flying through stars. While he was sitting on an inheritance, Musk wondered when humans were going to towards Mars. In reality Musk didn't just would like to visit Mars and explore it, but he also desired to establish a colony on it to ensure it was able to support humans.

He proposed an initiative he dubbed Mars Oasis. It was a plan to establish a greenhouse on the red planet and demonstrate that it can be utilized as a place to develop. The idea behind this project was to stimulate interest from the public about space travel, and to accelerate the advancement of technology for space.

In dismay at the lack of energy shown by the government in the efforts to develop space technology, Musk decided to take on the challenge by establishing a private company. Three times, Musk visited Russia to explore the possibility of buying rockets.

When the first time Musk was on the trip, he had a meeting with a handful of Russian companies in space exploration, however was treated with disrespect by their chief design team. In the end, the trip failed and Musk returned empty handed. In his subsequent trips he sought to buy ICBMs with no warheads. He visited Kosmotras, the company that makes them Kosmotras once more to discuss the possibility of purchasing missiles at 8 million dollars each.

It was not the way Musk was looking to do. Instead Musk quit the meeting quickly and then began planning the future direction of his company. It wasn't a vision Musk would ever forget. As he considered the failed agreement, Musk saw that the problem with space travel was not the technology available and the techniques haven't advanced in the past since 1960. Instead of relying on high prices for outdated technology, he decided that he could do a better job of it himself. Based on his calculations made during his return flight to America and back, the cost of the basic materials needed for making a rocket could be 3percent of the cost the Russians were seeking.

In 2002, Musk established SpaceX and set out to compete directly with the world's governments in the last frontier. At the start of the business, Musk raised over $100 million to achieve the dream.

As soon as he got the launch underway, Musk worked to reduce the production costs, specifically in the engines and electronic components for the launch. As usual, he had to make sure that the company was small and agile to be able to perform even in challenging times. By cutting down the bureaucratic weight that most businesses carry and refining the technology in tiny ways over and over, Musk was able to make the company efficient and sustainable.

If asked about the way he learned about in rocket sciences, his response was also in line with his experience. Musk sat in silence and said in a serious tone that Musk "reads many books." As with programming, and just like his qualifications, Musk was able to get the information he needed by doing his own research.

The aim was to build the ship reusable and cut the cost of getting to orbit. When SpaceX entered space, direction of orbit's cost was actually worsening and was in the early 1960s. SpaceX was established to change this trend, allowing people to continue to travel in space and the stars above.

Musk is extremely enthusiastic about a time when humanity is an interplanetary race. Musk believes that a future in which space travel becomes accepted as normal is an exciting prospect and that if humanity did not achieve this goal that could be a very tragic thing. It doesn't mean that he believes there should be a an obsession in the human race to invest as much as they can to fund the space exploration project, but humanity should invest an amount of money to fund the advancement in space exploration. It's not required to be funded as the war effort, but maybe, it is more important than the ice cream.

The approach he employs this time is similar to the leap to design Zip2 software, a program that will let newspapers build websites and eventually PayPal which is a service that connects people to each other. Musk is the first to get the funds to develop technology and then makes them available to the general public once they become cost-effective. It's like Robin Hood entrepreneurialism.

SpaceX required him to invest huge chunks of capital investment. Musk's plan was to have three attempts to launch rockets into space. in the event that none of them worked and the company didn't succeed, it was not going to continue. After having named the rocket Falcon-1 through a Memorandum about that of the Millennium Falcon from Star Wars, Musk set to work tirelessly.

In the year 2006 Musk made his debut launch. The launch was not successful, but Musk was determined in his goals. A lot of money was burned and he was also facing many other challenges in his personal life including a failing marriage and the struggles with his company Tesla that was created in 2003. Musk was in some of the toughest and difficult periods of his life.

Flight Two performed better than Flight One. It did manage to make it to space, but didn't make it to the orbit. This was not his last attempt since he had intended to make three times to get into space, however the business seemed to be somewhat of an odd idea. Launches were not making much use of the spacecraft.

However, flight three did not the success was hoped for. It took off, and was as successful as flight twoin launching into orbit, yet it did not reach orbit. The task was not easy and as Musk states it is about 70 times more difficult to attain orbit than to go into space. To enter outer space, the rocket has to attain Mach 9. However, for it to be able to orbit, the rocket has to be moving at a minimum in excess of Mach 25. The distinction between the various degrees in Mach speed is the energy squared. This is the reason it is so difficult to attain higher levels.

Musk immediately made an appearance and announced it was working with problems with flight 3 and knew how to deal with these issues. Musk reached out to their investors, and they raised funds to launch a fourth flight. In the following months, they planned to attempt again, on September 28 the year 2008. This was unprecedented in the realm of space craft. Their timing was daring.

Chapter 5: The World By Being Sexually

Attractive

While he was determined to change the way space travel is conducted, Musk felt that there was more to be done. In addition to working towards preventing humans from destruction they could be able to spread throughout the world Musk was also determined to stop the destruction in the event that it was possible. One of the biggest areas of his attention is to reduce the dependence of humans of fossil fuels.

Musk recognized that there was no reason electronic vehicles couldn't be competitive in the marketplace with counterparts in the gasoline market. Being sustainable economically could come with a beautiful image for consumers and it doesn't have to be unique. There's no reason it couldn't be possible to have electronic vehicles that were pleasing to the eyes. Motivated by his love for cars that were exquisite, and dissatisfied with the look of electronic automobiles, Musk was hungry to create a market for alternatives to energy consumption.

Martin Eberhand, Ian Wright Martin Eberhand, Ian Wright Marc Tarpenning had a very similar concept. In 2003, they launched their own company Tesla Motors. With no money to construct their initial prototype, they had to find investors. In the early days, Musk was a well-known persona of Silicon Valley, a massive success. Musk was the name you had to be aware of if you wanted to participate in the opportunities Silicon Valley had to offer.

Not only Musk Musk interested in investing into Tesla Motors, writing out the sum of $6.5 million and a half, but Musk was also able to think for how to make the idea become a realisation. He was the most significant investor in the company and through this acquisition that he was appointed Chairman of Tesla Motors. From that point on it was his intention to develop a deep connection with Tesla's design.

It is the very first American automaker after the launch of Jeep began in 1941. There was no one else who believed there was room for a car on the market. Musk was convinced in his belief that Silicon Valley could provide something that Detroit did not have, and that was through the use of batteries, electronics as well as electric motors. With his knowledge of technology, he felt that there was a path to commercializing the electronic vehicle. There was nothing similar to it in the Midwest and industrialization was slow. At the height of the automotive industry Detroit firms would complete an automobile in one place, put all of the pieces on a truck, then drive these pieces to another plant located in the town, then then continue the assembly. This was not an innovation in technology due to its streamlined nature and efficient, but due to the quality of design and the emergence of new technologies. Musk sought to draw the inspiration of this and marry it to an application that could result in efficient vehicles that utilize the latest technologies that are the hallmark of modern technology.

Tesla's concept was to make an affordable low-cost vehicle and a mid-cost medium-volume car, as well as a cheap high-volume vehicle. It would be able to satisfy all types of customers in the market, and allow electronic vehicles to be a part of the various classes of cars. The first line they would create would be the highest-priced line, as there has ever been an electronic car so powerful and attractive as the top gas-powered automobiles.

Parts of the original designs were bought from different businesses, for example, The chassis was bought from Lotus. They were used to make the car however, they eventually produced defects in the design, such as the size of the chassis was too big. Musk likened the process to building a new house. is, when you put many hours into building it is better to have leveled it up and started new. But, with a group of 18 people, the design completed in October 2004 and completed in the early part of 2005.

The prototype was a hit with the car-loving millionaire. Musk tried it out and found it to be satisfactory as the first step to realizing his dream of having a desired electronic car. Musk invested another $9 million into the company , and then went around seeking funds to fund the venture.

This model was eventually renamed it's Tesla Roadster, which was shown on television. The first individuals to test it was the then California governor Arnold Schwarzenegger, who drove the Roadster in the year 2006. He was awed by the car and was one of the first individuals to buy the car. George Clooney was another of Tesla's initial customers.

However, as the business increased, things did not work like they did. Eberhand had played a significant role in the company when it was small, however, as the company grew and expanded, the company required different things. The competition only increased in the automobile industry, so Tesla had to operate with no friction and at a cost as low in the way they were able to. The costs of producing automotive batteries added tension.

Musk wanted to begin using Model S. Model S, which was intended to be a medium-cost and medium-volume model. It was intended to be an affordable family car, but Musk wanted it to be attractive. Musk employed Henrik Fisker to help with the designs for Model S. Model S.

Fisker's designs were terrible. He had designed a car that was a disaster and refused to work on behalf of the firm. What Musk did not know was that Fisker was working on behalf of his own firm, Fisker Coachbuild. Fisker was a fugitive from the business and had originally planned Model S. Fisker had resigned from the company with plans for Model S and announced that the company was planning to launch the Model S in 2010 -- the year that Tesla was planning to launch their own model. Tesla was eventually forced to file a lawsuit in April 2008.

At this period of conflict that Musk was able to hold an emergency meeting of the board and spoke directly to the company. Musk wanted to be informed of what was happening as well as how things were managed, and if there was any issue. "If you're not satisfied go through a divorce," said the man at the time of the meeting.

At the close of the meeting Musk was able to remove the title that was the CEO position from Eberhand. It was Musk who would become the CEO and take on complete charge of the direction for the business. Eberhand further aggravated Musk's problems by filing a lawsuit, claiming that he was dismissed as CEO due to the fact that he was used as a reason to blame the company's failures.

No matter the fault of whomever, Tesla was facing difficulties. In the blink of an eye the 2008 financial crisis was taking shape. The company had just completed the first prototype of their luxury vehicle three years prior which meant that they were not in a position to create a viable model. In focusing on the wealthy the company put itself in a precarious spot in a time when there was no money to spend. There was no lower option and the Roadster was priced at $109,000.

At this moment, Musk was being drained of all things. The media was constantly hounding Musk for his failures with SpaceX and an ill-fated relationship was forming between him and his wife whom he divorced in the month of June of 2008, dissolving the family into five children. Justine Wilson, his ex-wife was later to make use of her writing skills to talk about Musk. She created a picture of him as a solitary and obsessive worker, barely capable of functioning as an affluent father or husband because of his total immersion into his business.

The perceptions of other people's opinions can't be managed, but that doesn't mean that Musk was able to let go of it. Musk was not one to dig into his relationships with others however, after Top Gear released a negative review of the Roadster, Musk sued them to stop them from airing the show on TV. The show stated that the Roadster would cease to exist after about 50 odd miles, yet it offered excellent handling and an attractive design. Musk declared on the grounds that it was fake and claimed that the Roadster will definitely be capable of completing 200 miles in a single charge. The case of Musk with the BBC was rejected.

The month of May, 2018 Fisker submitted a request for arbitral proceedings in the lawsuit Tesla was launching on behalf of his business. This was a case with many implications for Tesla since they both Model S and the Fisker Karma were intended to be hybrid vehicles that were both powered by gas engines , which were powered by a generator that powered batteries to generate electric motors. These designs made them competitive, even if they weren't as an imitation.

In November of 2008 Fisker was successful in the case and the judge ruled that he did nothing wrong. The court said there was enough space in the market for both vehicles. The court also required Tesla to make $1.14 million dollars in legal fees as well as a reward to Fisker. This was a total discredit to the work of Tesla.

When he was reflecting on this period, Musk described 2008 as the most difficult time of his existence. Musk does not consider himself to be an individual who would suffer from a depression, however that if he did suffer from one it was this year that to it. The entire world was in turmoil and his prospects for the future were uncertain.

Yet, Musk was never the person to lose his determination. He was more devoted to his money and invested every penny he could into the business. The entire time could have been used for Tesla and efforts to keep the business viable. Many critics attended for the view of watching the business lose money as it spent ever more money in order to keep it on the right track.

In December 2008, the company's prospects weren't promising. At Christmas, the firm was only days away from declaring bankruptcy. Musk was sitting on his own at the time, his family and his accomplishments ahead of him, the uncertain nature of space travel and a risky entry in the automotive industry during the time that General Motors went bankrupt ahead. There was no way to know what was in store or what Musk could accomplish.

Everything that wasn't strictly vital was shut down. Cost was not the one factor that was in charge, If it didn't earn money, then it wouldn't get completed. Tesla was beginning to look like an increasingly unlikely idea. There was only a week of financing in the event that Musk could stretch it.

Chapter 6: Thoughts Coming From A Festival

When he was at Burning Man, Musk was in a state of trance. The dependence of the world on non-renewable resources was a problem for Musk. There had to be a better solution to the future to allow people to access energy that was available to them in a way that was safe easy to access, as well as easy to access. After contemplating it for some time, the idea was born. He could engage people by their environment.

After having the time to take on other obligations, Musk pitched the idea of SolarCity to his friends and offered to finance the venture. The idea was to take the resources accessible to everyone every day, and transform it into a source of power. In the event that Tesla is responsible for the consumption of energy and production, SolarCity was supposed to be in charge of the production of energy. Musk was the only chairman of SolarCity however the Rive

Brothers Peter as well as Lyndon Rive were behind the tasks required to run the business. The construction of a solar field with panels can be expensive as is purchasing an entire field in California. Instead of investing in huge areas for technologies for solar, the company decided to build rooftops. In focusing on the neighborhoods of California they were able to gain the space needed for establishing their tech. The problem was that the panels were expensive, which meant they would be difficult to convince a consumer to buy.

SolarCity has gotten around this issue by offering the panels no cost. Instead of having to pay to the business, SolarCity would install the solar panels, and later provide the energy to homeowners to use in order to reduce the cost of their electricity. This meant that the customer will benefit but would have to not take any risks, and SolarCity could keep ownership of the panels they would spread.

Furthermore they followed the design of Musk and Musk, where the objective isn't to improve the product by a huge amount however, to create an outstanding product by making various small improvements that work to produce a sleek and seamless product. SolarCity has been able to create efficient panels that weren't overly costly. They gave a crisp look to homes with a reflective futuristic feel to roofs, and showing curiosity about the natural world and the ability to be a thrifty.

The strategy was extremely effective for the company and within a short time, they were the leading solar energy provider in America. The majority of their customers were located in their region of the West Coast, California to be precise. They were a hit, successful and grew quicker than any of their competitors.

Many large companies made the decision to purchase panels through SolarCity as did eBay. The purchase that eBay provided SolarCity was the biggest solar panel installation ever in San Jose, California. In the identical year SolarCity was able to complete the most extensive solar installation San Francisco for British Car Motor Distributors just two months after collaborating with eBay.

SolarCity was making its reputation as the leading solar energy provider. Its success was not expected, firstly because alternative energy was an obscure market, and the unusual leasing model was in place, which demanded consumers to pay nothing upon the installation.

This created difficulties that were a problem for SolarCity however, since it was a very costly business model. The company began losing money because they had put in excessive panels without earning any income immediately. Their business model was based on the long term and eventually SolarCity could be one of the top energy companies in America through the way many homeowners have signed up for their solar panels. However, getting there requires a substantial amount of money to finance. Initially, SolarCity decided to put the burden on themselves however there was a debate about whether the money should be transferred to the customer.

The bank that was in support of SolarCity has pulled out of the agreement and left the company floundering. The long-term plan wouldn't be able to implemented in the event that they failed to survive. This was happening in 2008, as well as other issues Musk was dealing with. Nobody could have anticipated the hardships that would come in the economic downturn, and the firms were not ready.

Beforethat, Musk was fresh off the phenomenal achievement of the sale of PayPal and was able to think about the future. SolarCity did not need to make money however, it had to be an option for energy. When the recession hit the financial market was critical. There was no time to take action that wasn't essential. At the time of this crisis SolarCity wasn't the one to be the center of attention because Musk was the only chairman of the company. However, the difficulties in finding success was weighing on Musk.

Chapter 7: Flirting To The Second

The launch of September in September flight 4, was extremely nervous. Musk has secured a new round of capital which he was not sure that he could receive, and the entire whole world watched. As his personal problems were discussed in the media and the difficulties of selling a luxurious car to a struggling consumer, Musk was tense. But his passion for his work was never questioned and he stood by his convictions.

Flight four was an enormous success, far beyond what Musk as well as SpaceX could have ever imagined. The spacecraft, equipped with cameras, climbed into orbit and proudly displayed before all the globe as being the very first private spacecraft to ever go into space. It was a milestone moment for Musk's business, Musk, and the world at large. The world was aware of his accomplishment, but this didn't translate into money or business.

The weekend prior to christmas, Tesla was close to being completely insolvent. Within a few days, it could be forced declared bankruptcy in the event that it did not get more investment. Musk will either need to use reserve capital or let the business fail right in front of his eyes. The company's fortunes began to change, and people began to doubt the company's capacity to cope with the demands that the marketplace has to offer.

Furthermore the debts incurred by SolarCity did not help in any way, and the accomplishments of SpaceX were costly. Musk's plans were placed at risk during the 2008 financial crisis. The first time in his entire life Musk was devastated by stress. Everything that was to come was dependent on the immediate actions he took.

In the course of this year One of the only things Musk was able to enjoy in his life, that was not directly related to his difficulties, was his friendship to Talulah Riley. They came across each other through an accident, Riley was unaware of Musk prior to that. She was not familiar with Tesla, SpaceX, and the rest of his professional history. This was a source of inspiration for Musk. After having lost his relationships and enduring the most stressful time of his entire life, the man required something to keep his sense of sanity. If not, the stress could be too overwhelming.

Musk's strategy of diversification is both an incredible strategy in the current environment and also one of the most risky strategies. Rome collapsed due to spreading itself too thinly. So by putting too much pressure on a limited resource every organization is at risk of being in the exact same situation. However, Rome lasted longer than the majority of countries and its effects remain evident to this day.

But it was not just that Musk have his interests divided between different companies, but he also was also a victim of his own interests between the businesses. SolarCity would not get any of his attention, while other companies were left behind in their goals. Musk was determined to take us to Mars in order to colonize Mars and be able to move from Earth, Mars, and further. He needed to build his own rockets as it was not a NASA interested in reaching Mars. Tesla was looking to expand in three different ways, an expensive model for enjoyment and art demonstrating that the car was able to function as a race car and a moderate-cost option for families who still had the extra features as well as a low-cost version to make it affordable and accessible. The goal was to provide every consumer with a cost-effective and efficient vehicle, thereby helping consumers to reduce the dependence upon fossil fuels. In the end, Musk was left with an electronic sports car that couldn't move across California and plans for a second car that was stolen by Fisker. Musk's ambitions were always high however sometimes ambition can get behind the advancement of technology.

Musk ended up not making an error of a magnitude; he simply didn't know how the future economy would unfold. His strategies were sound and they appeared like they could work however, Musk did not have numerous economic recessions in his entire life. At the time Musk established the first business the economic outlook was good, particularly in the technology sector. If things had continued like they did, everything would have been going well. Musk has a remarkable ability to realize his ideas.

It's not always about application of understanding, but. Even with his ability visualize the world and plan out what he would like to accomplish, Musk wasn't being flexible enough. He was working the problem as a math or physics problem , whereas the issues that face the world economy have more like an art. To be successful in the global economy you have to build alongside, with or on others. If there aren't opportunities to build in the first place, you will not be able to advance regardless of how impressive your knowledge. Even if you can make a beautiful and entirely environmentally-efficient building for a decent cost, you will struggle to find customers during a housing crisis. Without the interest of other people it is simply not an possibility to showcase your talents. The market will always determine whether something is worth it.

This implies that life can't be solved by quality. This was a mistake made by Musk. While he was able to do things right, and was pushing forward into new ground but if there was no demand for his product, he wouldn't have an idea for a business. The idea is to utilize all opportunities that are available to the best way to raise capital. this is a fantastic method, but it relies on having something to begin with.

What's essential to remember is the risk of chaos. You require the tools to help you create even in the most difficult of times. Don't take things lightly and discover ways to keep your efforts and yourself effective enough to survive with a limited amount of resources for a long time and then increase your stock of funds. If you turn this into the pursuit of building funds, in addition to taking on risks and maintaining a reserve then you can play the game with much more security.

Musk was over-achievers if there was anything. The company he was working for built rockets, with one of them that was aimed at creating various types of vehicles, and was also tied to a business financially, which was securing California and the remainder on the West Coast with solar panels. Each of these ventures was worth it and Musk was able to fund these projects, however because he was invested in them all simultaneously, he put himself open to a lot of risk. Instead of focusing exclusively on Tesla and possibly having two models available He had a lengthy collection of accomplishments without really accessible consumer goods.

He was not as wealthy as he was in the past but he was well in the realm of a millionaire, but the fact that he was the primary investor in both firms resulted in capital shrinking quickly. There was a high amount of risk threatening Musk from a variety of angles.

Perhaps, had he displayed more discipline, Musk might not have been able to achieve his incredible achievements. On the 23rd of December NASA gave SpaceX an $1.6 billion agreement to replenish the ISS. The SpaceX founder didn't know what to say other than to utter "I am in love with the SpaceX crew!" His insistence on an efficient flight resulted in a successful flight and SpaceX was saved. It had secured its future thanks to flight four. It was a significant achievement and was one of the most significant events for the private sector in space technologies.

Risk is a key aspect of success. Musk did his job by extending himself at the right moments and took risks which led to his place as a figure in American history. Another alternative is to fail completely. If you wish to achieve to achieve the same level of success as the people who make the world, you'll have to risk everything. If you're seeking success in your own life and make interesting contributions, make a decision to take chances. They don't have to be massive however, by putting your money on the line you can achieve great success in the event that you can make the most of this opportunities.

It all depends on the person you would like to be. If you're looking to be a subject of a book and be a part of the news, then you'll endure the long-standing Chinese curse of living entertaining lives. Anyone who has had a life worth writing stories of has ever had a bad time getting there. There are a lot of people who are successful but boring however. They're not as fascinating but they're steady and they know what their dinner plans are likely to be each night.

However, despite the achievement, Tesla was burning money. SpaceX demonstrated that it had potential for growth in the future and would benefit Musk's investment and effort but the contract wasn't enough to save the entire company. SolarCity as well as Tesla were in danger to be destroyed.

In playing the high-risk game that Musk was, he diversified his interest and potential. Although it's not advisable to put everything you own into your companies, this gives you access to an enormous amount of work if attracted to this. This meant there was always something to be processed however, it also meant he was putting eggs in multiple baskets.

The end result is that he would be able to escape from SpaceX. SpaceX. As long as he owned businesses that had been successful, Musk would have enough to pursue his interests however, Musk desired to be successful across the board. The companies were not founded to earn profits, Musk had enough money but he was looking to advance. The idea was to create the future of humanity. He wanted to think about the future from multiple angles.

This meant that Musk could play multiple games simultaneously with wins and losses in a continuous stream. Musk has stated that he's lost many battles, but not a conflict. Through advancing on multiple objectives the success of one area can bring relief to another. In this way one can observe that Musk was operating in a situation that was characterized by high risk. The majority of his money was invested in projects. However, this also meant that once a successful event occurs, the overall situation improves and the success is repeated.

The most risky part of long-term investments is staying to the same investment for long periods of time. There are a lot of businesses that failed, but only need to stay for a few months. Musk's story is unique due to the fact that he's an entrepreneur and genius however, there are also the lessons to be learned from this.

Insisting on investing everything in your various passions can lead to massive influence. There are projects that are creative or as effective as you would like to be, however it comes with the possibility of losing all of your assets. These investments will work harmonious with each other However, success at one level can result in success on another level. The more you desire and can successfully manage risk as well, the more you'll be viewed as a person capable of doing it. This will build trust and the chance to build. Another lesson concerns thrift. If Musk was not able to keep costs at a minimum across his various companies then they'd be forced the option of declaring bankruptcy. SpaceX was not founded simply because Musk would like to create rocket vessels and rockets, but rather because he worked out that he could build the rockets for less than just 3% of the price of purchasing one. The primary goal of the business is to make space exploration practical and economically feasible.

Keeping budgets in mind and in the service of the customer, whether it are the typical person or NASA can be an effective method. The most essential factors in earning money is saving money. If you are able to do this, then you will be able to balance a business.

So, with his reputation and his future were secure. Musk can rest at ease however he had to talk to Tesla. There were a lot of worries in the office and many people suggested that investors buy stock.

It was also the Roadster The Model S, the Roadster Model S, and the smart car batteries that were being developed. Three projects to manage and everything else that was not needed was to be taken down. Musk would phone Maye in the course of the year, and she'd have no idea what to say to him or how to respond. It was a lot happening with Elon and his family, so many problems, like devastating realities. It was a harrowing experience to watch her son suffer through this struggle , even though Elon had already proved his worth as an businessman. For her, comfort was found in listening to Elon speak about Talulah Riley. She was just beginning, didn't know much about Musk's work, but Elon was convinced that she was a great person. This is what mothers want to hear.

ELON MSK TOP 20 rules for success
1: Be True to You
Nobody was aware of the extent of what Elon Musk was doing as an infant. Some kids believed he was a strange person or somewhere in between being a dork and freak. The world viewed him as an unproven dreamer. In between Sci-Fi and comics The young Musk realized that it was necessary to build a bridge that he'd have to construct with the universe surrounding him. As he matured and grew in his physique and his potential, Musk would build those bridges.

However, this was not enough to stop Musk from the name of his spacecraft Falcon 1 after the space production, Star Wars. Also, it didn't stop Musk from creating a company from a pun on The Boring Company. The pressure from other people was not enough to force Musk off his path He remained committed to his ideas and desire, so the entire world benefited.

There was a lack of belief in the potential of an electronic vehicle however, before that people were skeptical about the power of the internet. During the early days, the Musk brothers were forced to bargain to negotiate agreements with businesses to get the internet on the market. There was no doubt that humans would be able to make it to Mars in the present but we could.

There was a lack of support for Musk when he first started his journey because no one was able to see what he was seeing. That's why he's the perfect example for humanity, since he has discovered potentials that were accessible to us, but were not realized. Through preserving this technology, he's found one of the most powerful technologies within him.

If you are a person with an internal voice within you that can see things and has a sense that there are goals you have to do and work towards, then commit yourself to these goals. You may find that you have a sense which others aren't able to see. There's an endless possibility within the universe and your perception is just as valid as any other perspective. The only thing you have to do in order to tap into it is to understand how to manifest the idea of possibilities into reality. It's a daunting task however, with a little focus and patience, you'll be able to accomplish this easily. Humans are designed to learn.

2. Lean from Your Environment

When Musk was in the school of South Africa, he used his time to learn. Instead of partying drinking alcohol and smoking cigarettes He stayed in libraries and played computer games. It was through this use of the resources readily available that he discovered the details of Silicon Valley, and was first captivated by the culture of America in the sense of it being the place where anything is feasible, and as the place of opportunities.

This insight drove him to Canada to make progress towards the promised land. In Canada, he took advantage of the opportunity to locate a school to go to and there he met an individual he could love. The relationship ended with divorce, but it lasted Musk through the era of PayPal. It was a significant aspect of his life, and allowed the couple to have five children. While at the school, Queens University, Musk was also able to transfer onto University of Pennsylvania. University of Pennsylvania, gaining access to America.

The importance of this school was not for the degree he earned however, while they proved his abilities, but due to the networking that the student was able to create. He had great friendships in the college and gained experience the social scene in an American setting.

After graduation, Musk had the opportunity to attend Stanford where he enrolled. When he left for California Musk realized that the surrounding was more powerful than the university, and the school dropped him after having been enrolled for just two days. When he got out and about, he began to design Global Link, forming the connections he needed to establish his company and attract customers.

It was this setting where Musk had envisioned in South Africa, that would provide Musk the platform to develop his ideas. Also, Silicon Valley had access to technologies that Detroit was not even equipped with and even more. Because of his position and the experiences in this region, Musk was able to create SpaceX, Tesla, SolarCity, The Boring Company, and many more. It was his access to engineers, technology and the zeal for new technology that allowed Musk to develop into the person was he eventually was.

Your abilities will only grow to the extent that you access the resources needed to let them. A genius without a context is ineffective. If you're looking to fully realize your potential and make an impression in the global community, invest in the world surrounding you. Choose a location that has access to the resources you require. If you're looking to develop the next big thing in MIDI move to a place where you can access recordings of sound and a wealth of natural talent so that you can market it with sound that is rich and stunning. If you are looking to build boats, then you need to move to a port city near the coast, with an established trade market. The more land there is to build upon the more high you are able to build.

This is vital to your achievement. You must be aware of the environment you are in. What's worth billions of dollars in one place isn't going to be a success in another. Find out what interests you and take advantage of the past that is already present around you. Human history isn't a thing that's been created, it's an ongoing ever-changing event which is happening each day. Each of us has the capacity to participate in it. Use this right.

3. Apply Uncanny Strategies

Musk didn't intend to become a pioneer in the space sector. Musk was looking to finance the launch of a spacecraft to Mars. As he tried to find a way to achieve this however, he discovered that no one was interested in the same mission. After looking into this, Musk decided that he could repurpose old technology to finance himself with his missions. When he tried to purchase this technology, he realized just how costly it was, and that it wasn't worth his time to fund missions using outdated and expensive technology.

Then, Musk founded SpaceX instead. First time ever in the history of mankind, a private firm was created to develop technologies for space. The public was shocked by the concept however Musk had a plan in his head. Instead of paying attention to general norms of the industry and the opinions of the masses Musk opted for the irrational strategy of exploring space with his own investments.

In just a decade, SpaceX established its credibility as a firm by launching the world's first private spacecraft that reached orbit. They received an agreement from NASA who are the ones who run the most successful space programs exploration. They're also the only firm that has ever offered commercial flights to space.

The other programs he has developed have similar roots. They're brilliant ideas that have unlimited use. Tesla was just as ridiculous as SpaceX however, it's the top firm in the field of electronic car sales. These concepts may not seem viable at first however the fact that they're not getting the support of the public doesn't mean that they're not feasible. Ideas that are unusual are often the most lucrative, particularly in the case of ideas that are unusual however they can be utilized within the everyday world. Look for something you can see that hasn't been thought of before and then develop it yourself.

Most wonderful ideas, such as space travel, or the implementation of environmentally-friendly technologies, or the ethical rehabilitation of drug addicts exist through the efforts of people who cared about them and made the sacrifices necessary to make them possible. If you've got ideas that you like, you should take the time to make them happen. The more energy you put into them, the more likely they'll become a reality within the world around you. Everything great comes from great ideas, so if you're blessed with an ability to envision possibilities that could be amazing that do not exist, then nurture the vision.

Your perspective could prove beneficial. Nobody knew that an South African would come to Silicon Valley then flip the power, automobile and industry of space on their heads. The concept was impossible, but nevertheless, Musk came, and his innovations were evident everywhere where he traveled. We do not know which influencers will be the ones of the future because no one is born to beone, individuals live to be the leaders of the future. Take up the challenge and be among the people who shape the future of our generation. It's all it takes is determination and curiosity in your surroundings and the ability to grasp concepts that aren't obvious to be sensible. Recall what it means to imagine of a new future, and think of a better tomorrow. In any case the future will consist from whatever ideas we think of to construct therefore we should be able to build the things we'd like to be able to see.

4. Be firm in Your Direction

There were a lot of times when Musk might have caved under pressure or stress but he didn't. There were plenty of reasons to be worried but he did not allow it to control his life. Instead, he decided to do what he believed to be a good idea and put everything in to achieve it. Always determined, he walked straight to the brink of destruction, negotiated it in a controlled manner, eventually found his way back into the sunlight.

The 2008 holiday season could have been the final straw for Musk's businesses. He wasn't required to take on the risk of investing the remaining reserves capital into Tesla however, he did this, and prevented Tesla from having to take incredibly difficult decisions like layoffs and selling off. Deals with Daimler that was signed a few days after the deal could have helped save the company, however Musk's investment took place before the company was aware that the transaction was in the works. Whatever the case, whether it was the main reason the company was able to stay in business whether or not it demonstrated that Musk was committed to his businesses and believed in his working with all the resources was in his. It was a sign of a man of determination, someone whom employees could trust and who the public ought to keep an eye on. Musk was not a lark and he poured his life in his job.

This was a further example of how Musk's determination and determination were able to bring him great results. Following the loss of flight 3, he believed in the knowledge that Space X knew exactly what was not working, and they'd try again in the next few months. It was a pretty absurd announcement in the world of space because no one could work using time tables with this short of a timeframe. Musk did not give up and achieved success in flight four.

Life isn't always easy. No person is able to beat an arcade game on the first attempt. If you're determined to get skilled at something, If you are looking to accomplish some thing, then you have to be committed to it. It is not possible to float through life if you wish to leave nothing in the dust. Each legacy is accompanied by the tax on blood.

Musk's ability to stay focused on his task is remarkable. It's not something everybody can imbue from their own, but it's something everyone should try to replicate. It's a good thing to be committed particularly when it's an oath to your interests. The longer you put in on something, and the more you feel for it, the greater your effort will be worth it.

5: Learn to manage relationships.

It wasn't an easy task for Musk to understand the people within his life. As a child Elon felt a deep love about his dad's absence and wanted to offer his father the chance to improve his life. Elon wanted Errol to feel depressed or lonely, and so Elon decided to live in the same house with him. Elon later regretted this mistake and regretted that the family had been in the company of his mother as well as sister during this time of his life. It would probably be healthier for him, due to the way dad was.

The mistake Elon made , and one that many people have been exposed to is the fact that simply because you feel sorry for one doesn't mean you have any obligation to them. Sometime, we find ourselves being just pathetic even those you cherish dearly and desire only the most for. Sometimes, those we love slip into bad behavior and seem to be overwhelmed by it.

In this situation it is best to act what Musk did when he was a grown man and to create a distance between you and the bad guys. Musk is no longer connected to his father, and he manages his own life, achieving even in spite of the promises Errol made, stating that Elon would never do any effort to make anything out of himself.

Negative energy can be very efficient in destroying the human soul, particularly when the soul is involved in actions that are uncertain. If you really would like to be a catalyst for change, or even to announce an optimistic development in our future as a species and as a species, then you need to be mindful not to allow this to happen in your life.

Being a change agent demands a lot of your. All it takes is confidence. If you aren't able to justify your goal, what could you think of achieving them? Self-confidence is among the most effective tools an individual can use Therefore, you must cultivate it and defend from the negative influence other people who could negatively impact it.

6: Don't Be Afraid to Explore Places You've Never been to

There was no invitation for Musk into North America, he learned about it. He researched it, and learned about it due to how technology was developed within the region. It was not clear who who told that he should be interested in it, and no family members who lived in the region or had any family history from the area.

But, Musk was determined to relocate to North America before turning 20 and hasn't moved anywhere else since. Musk settled there. North America, he made millions in

North America, and he's made billions of dollars in North America. His sense of intuition was completely correct even though he had never traveled to the region prior to making the move. He was aware that it was his desire to be part of the work that was happening there, and so he went there and offered his services. He paid his fees and demonstrated that he was able to contribute.

Being a part of it, and achieving success in the area, is also what led him to the opportunity to be part of Tesla. It was because of his fame as an exemplary Silicon Valley success that he was approached by Tesla's original Tesla founders. While he was there and became one of the characters were would read about as a kid and was one of the people working to advance technological advancement in the promised technology-driven future.

These kinds of predictions are vital. Everyone would like to relocate into Los Angeles, but few would like to take a shot at for the Asian Carp out of St. Louis. If you are motivated to preserve the ecology in the Great Lakes, and for those who want to see an industry return from the Midwest There are numerous possibilities to relocate into St. Louis and set up fishing ventures. Fish aren't well-known in America however there is a market in the world for the fish. A skilled fisherman who has an interest in seasoning may also attract the attention of the American publictoo.

There's an endless amount of opportunities across the globe. Everybody has a skill set and there's inspiration all over the place. Find a way to contribute to that. Find a cause that will be benefited by your contribution or a place that is worthy of your contribution and then take action. If you need to relocate or disrupt your life, move, especially when you're young. There is always a chance however not all opportunities is available from all over the world. There's always a moment and an opportunity. Find out that and develop a sense of working with systems that are not part of your own.

If you're able to do this, you'll be able recognize opportunities that are available to you and your talents. Don't be afraid of following the direction you feel. There's a reason you're drawn to what you're attracted by. As you become more committed to these emotions, the more you contribute to the world you'd like to live in, the more likely you are to become a participant in it. Once you're member of the group, you are a part to it.

7. Don't Be Afraid to become something that you've never seen before

There are few as famous as Elon Musk. As he grew older, he didn't have many teachers who could be a positive influence to his inner psyche, particularly due to the amount of resentment his father felt for his son. That meant Elon was forced to develop his own strengths with no an example in front of him on how the use of these would unfold.

A lot of the family did not be able to figure out what to make of him too. The fact that he had a photographic memory of a lifetime and an desire to learn left Musk alone, even with his mother. Musk was an independent man. His ability to comprehend grew as a child and never slowed down.

This gift brought greater opportunities to the man as the world progressed. It started out leaving him in a new nation, an immigrant from America living in a small office in California while he attempted to achieve his dream for the advancement online. Very few were brave enough to venture into anything of this kind while he was in the wild, so he left the Ph.D. to follow in order to explore the possibility of establishing an online business.

He didn't know what the company's future would be like or what the future could have in store, but he developed an application and an sales team. Similar to his success at the age of 12 when he bought an online game for $500 There were no rules for Musk as there was only the realization of the potential.

By pursuing his ideas, Musk was able to establish a company and later make it millionaire for the very only time in his lifetime. Nobody could be critical of Musk since no one had anyone similar to Musk. Musk is truly an explorer; he is always looking in the future and often overlooks the present.

The only thing Musk was looking for to guide him in this direction included books written by authors Musk loved, films that loved him comic books, his idols like Tesla as well as Neil Armstrong. There was no one who could give his trust and his thoughts. Musk was required come up with his ideas and his strategies by himself.

This has made him Musk being one of the most powerful individuals in the contemporary world. Through his own reflections, and by bringing life to his own vision about the universe, Musk has become one of the most powerful men to has ever existed. Even without a dad, it is possible to become an outstanding man. The most important thing is that you make a commitment to yourself. Once you have done that, anything can be accomplished. When you have the confidence of yourself, and the confidence of yourself, you are able to accomplish whatever you desire to.

Set your own standards; make your own unique character to be. In doing this you'll create your own self-image that you can display to the world without anxiety. If you choose to do this the perception you have of yourself will not be restricted by the prejudices and limitations of the people who have come before you. Take a look at the world through the eyes of a child and you'll be able to see errors that the world of the past doesn't have to be concerned about, however you'll also be able to access many perspectives that they've been unable to encounter. Being innocent is risky since you'll be exposed to dangers and danger, but if your ability to take it, your innocence will be experienced rapidly, and you will be a powerful version of yourself.

8: Keep Your Vision

Your desires aren't "nothing". If Musk was dreaming about spaceships when he was a child the idea was considered typical of young children, and not an passion for traveling to space from another planet. But it was this dream, that idea that inspired him to create the first company offering personal space journeys. In the present, he's establishing businesses to research and develop cerebral implants as a way to transfer processing and memory of humans to a technology. He's also pushing for the advancement of artificial intelligence with the hope of a better future and avoiding the Terminator scenario. The man has set the goal of digging tunnels underneath LA in addition to New York City so that the traffic won't be so terrible.

These are all the dreams of a child. It's like they were thought up in between Science classes and English but they're the plans Musk is extremely committed to and implementing. There is nothing he will allow to hinder the achievement of some of these objectives and he'll be able to achieve them all provided that his health and the world permit.

This passion for interests similar to those of a child drives Musk. Musk doesn't wish to improve the efficiency of an engine, he wants to make it more efficient, but he would like to transform the world. He is in favor of renewable energy, he's looking for ethical cyborgs, electronic vehicles as well as space transportation. This dream, this vision of a brighter future for humanity is what drives Musk to pursue every field of fascination to the point that his ideas might not be as absurd as they used to be. Musk has invested the funds and the time needed to develop these areas, and has made significant progress across all of them. Musk recognized a path no one else would be willing to follow and, instead of dismissing it to be a dream or being naive, Musk embraced the challenge of his ambitions and took off as quickly as possible.

Nobody knows the mechanism of human imagination. It is possible to conclude that it's an action that is driven from the cumulative accumulation knowledge which creates thoughts and images that are a result of the information you've seen and heard. This is a way of defining and a concept of how the mind functions fundamentally, but in the end, it's not a satisfactory explanation. If that's how the mind functions in the first place, how can one explain intuition? What are the ways to explain the concept of religion? Is it possible to explain a desire for things that aren't rooted by experiences? Can we be sure that we are able to only perceive the time with one eye?

There are many concerns about consciousness and they are difficult to be answered. It's all about faith. If we can admit that we're not sure what the mind does maybe we need to think about our interests and desires. There is no way to know what drives people to pursue the passions in the way they do. But there's something that's going on. If you're a fan of marine vehicles and have been for the last four, perhaps there's an idea that you're destined to develop that will transform the way that ships travel. Maybe this is a result of a coming event, or your design of plans. Maybe fate is calling you. Whatever the reason, there's something that we cannot say the exact nature of it. It's crucial to be aware of your own vision. The goal is to demonstrate something. All you need is to be willing to hear.

9 Keep Things Amiable

It's easy to expect Musk to be angry and angry after being dismissed as CEO of PayPal however, he chose to not pay attention to the anger. He was dissatisfied with the motives the board used for the removal of his position from him, however he wasn't going to force his way into them during an important time for the development of the business, nor would he wish to provoke anger. He was aware that it was more beneficial to have more than fewer friends.

A business relationship is an extremely important thing, particularly in the event that they're interested in investing into your venture. Musk was recognized for his good manners of not getting emotionally involved with former colleagues, because after he establish his own companies following his sale to PayPal many of them came forward to invest in his new venture. Peter Thiel, who went to succeed Musk as the CEO of PayPal was one of early investors of SpaceX. Although its internal operation of PayPal wasn't without its own drama but the relationships that had been created there had a lot of utility or the need to serve. The removal from the position of leadership wasn't a big loss for Musk however, he was the biggest beneficiary of the earnings generated by the sales of PayPal and also kept his strong connections to the business.

The value of emotions is not more than advancement as well as business acumen. If you have to decide between emotional release and keeping the possibility of economic growth The best choice is to preserve the potential to earn a profit. Don't take yourself to the ground in order to satisfy your desires There are more feelings than feet.

It may be difficult right now however, it will pay off in the end. The less enemies you create more friends, the better off you'll be. There are exceptions to this, for instance, Internet trolls and media detractors All successes are likely to result in these. It is important that you don't make enemies of your relationships. If someone has had a relationship with you previously or gone extra mile to assist you, be aware of their issues. If you have someone who has worked with you before is likely to be more than happy to work with you in the future, particularly when they have seen you succeed the first time around.

Friendships aren't friendships in the business world, and if you're looking to work in the field and work for yourself, you must keep your mind off the work you do. There's a private life, and professional life. Separate them for the success you deserve. The reason is that you may not like someone who could be your biggest financial asset. It is not possible to have a good relationship with every acquaintance or business partner but you'll be able to get along with some individuals who either harm your company, or are incompetent enough that they undermine the value of your job. It's not about choosing which people you'll get along with and work with people you are comfortable with. In the end, Thiel and Musk had a hugely successful time with the acquisition of PayPal which is why it was fairly obvious that why they could not continue working together. Their track record proves that it was a successful undertaking.

By keeping things private by keeping things unpersonal, you can keep access to resources and keep your business in a stable position.

As a business entity it is essential to be able to do this. The more opportunities available to you, the more secure your growth will be. That is how you can be successful as an entrepreneur. even musicians can benefit from this experience. The effects of emotions can destroy a lot of potential from music to business. When stability is established in relationships and life and relationships, potential can be channeled and utilized. If emotional turmoil continues to ravage your life and ruins your day, it's difficult to conduct business.

10. Live for the Future, Not just for Fun.

Despite the fact that he purchased McLaren F1 despite the purchase of McLaren F1 after the sale of Zip2, Musk has always been a smart businessperson. Musk isn't the type of person to spend money on unnecessary expenses because it's not efficient. It's not doing anything. Instead of purchasing a house using the money he made and wasting his time in Hollywood seeking an older wife, Musk decided to use his wealth to create companies.

Before that, following Zip2's success, Zip2, Musk was quick to finance another company with the capital , and then get to work. The result was a business that would later develop into PayPal. Through this constant dedication toward the future Musk is able to achieve success after success.

It is possible to live life to be enjoyed in the first place, however, the foundation of life will never be laid. The time spent enjoying yourself is typically wasted time The true value of having fun is the effort you do to gain it. It's much more enjoyable to go out drinking when you've got something to drink, and perhaps, something you can celebrate.

Do not forget to enjoy yourself However, be aware of how you use your time. Find out how much time is just passing through the time in anticipation of something exciting to occur. The time you spend waiting is a wealth of resources that you can use for everything. If you were to play the game of beer pong long enough you'd be able to make a couple of dollars off it. The idea is that talent can be accumulated but the most important thing is reputation. As you gain experience and gain experience, you'll be more likely to get opportunities due to your skill as well as because that of the reputation you've earned.

There are a few people who choose to invest in the future over the enjoyment that they are having right now. This is an extremely appealing aspect, and if you are able to demonstrate yourself as someone who's got something happening in the next couple of years, people will think they're that person. If you can prove this, you'll be able to gain being able to count on the confidence of other people as well as the opportunity to develop your own ideas in the perspective of the trust you've been granted.

Musk could have purchased an island chain; Instead, he put his life on the line to become an influential voice to shape the next century. The result of this is earned him a spot in the ranks of pioneers of America and the field of technology overall. This prioritization of work over pleasure which has brought him this great success. The people who know him personally might consider him to be a bit obsessed with his work however it's difficult to deny such impressive outcomes.

11. Know How to show that you're a Human Being Too

In the year 2018, Elon Musk appeared on the Joe Rogan Experience. It was an extremely memorable moment for Musk and host. It was, in fact, one of the top personalities who has been on the show, which is why it was a turning point in the career for Rogan. Through the interview, Rogan demonstrated that he was not just an entertainment show, and more than having conversations in a casual manner with people from contemporary society. The interview with Musk demonstrated that Rogan could connect with the influential like any other anchor on TV could. Rogan could also offer the audience with a different setting and provide hours of conversations instead of a ten minute segment. This wasn't just crucial in Rogan's career but it was also a pivotal event in the whole field of alternative media as it demonstrated that there wasn't a have to be a middleman between the influencers and the general public who would interpret events in a particular manner. The simple style of alternative media was a major success and proved that even the most insular people in the world was able to be understood without filter.

For Musk this event was a huge one too. It wasn't just because it was his few interviews. Musk is a well-known name for his line of work. However, most of his appearances are in high-end environments. When he is interviewed the interview is not of him, but one of the legend. The interviews focus on his space exploration as well as his current projects, his stock in Tesla or his personal story. They are all respectable for a man with his stature, but there are occasions in which Musk smiles, as when he's arguing with a reporter.

The Musk who was featured on the podcast of Rogan was totally different. In lieu of talking from one suit and another dressed in a stylish and tailored ensemble, Musk actually underdressed compared to Rogan. Rogan hosted the show in a soft pink dress shirt that was unbuttoned on the collar. In contrast, Musk appeared dressed in an unisex black shirt with "Occupy Mars" written on it "Occupy Mars" on the front. The atmosphere immediately changed and the two people appeared as one although Rogan was in awe present with Musk.

The podcast continued and the two enjoyed whiskey and then talked about more. The discussion covered everything from the development of AI and whether the technology's development could be a threat to humans and if there is any measures that could be taken to stop developments in the event that they prove to be negative for humankind, whether or no aliens are on the horizon to more. This wasn't your typical interview Musk was given. Musk was asked to give, instead it was a video interview that saw Musk was smoking marijuana for the first time in the few moments of his life live online. In a short time, Musk was laughing when the phone was checked to see his pals' shock at his actions.

The podcast's video has received around 20 million people who have watched it. The video went viral and revealed the impression about Musk to the world that was more transparent and more likable than anything that had been published on Musk before. The popularity of the video might be diminished by those who claim that it had an adverse effect on his image. However, it was only affecting people who wanted to denigrate Musk. For the remainder of the population it revealed the human side of an entrepreneur, who is a mystery. For Rogan the episode on his show is the most-loved episodes ever broadcast.

Whatever you accomplish, you're human. Being aware of this, respecting the fact and staying within the boundaries of your body will lead to greater things. If Musk did not like speedy cars comics spaceships, spaceships or conspiracy theories What else would you have to keep as an individual? His fundamental characteristics shape his reputation because he's accomplished such feats of inhumanity that people are eager to learn more about the person behind him.

The public's exposure of his character public does not pose any risk to Musk. In the final analysis, he's an eccentric and geeky person who is concerned about the planet. The general public is generally receptive to Musk. Through the ability to display his personality in settings such as that of the Joe Rogan Experience, Musk shows that he's living also, that he's able to be a good listener and enjoys interacting with those that aren't directly beneficial to Musk.

This is a valuable lesson, as achievement isn't appealing by itself or even great. It requires humility to share itand the an attitude of humility to take pleasure in it. Take a break from your horse if it's on and go for a walk in the dirt with everyone else. It will make you appear better when you return up, because we'll be able to tell how you look when you have a little dirt on your skin. It will help the people around you to trust you when you can unwind and display the side of you that is there in the absence of trying to present yourself in a way that is acceptable.

If you share a genuine version of yourself and being able to be comfortable with others, you'll be the most effective advertising for your business that you can be able to secure. Be true to who you are, and share that with others and be authentic. You'll be rewarded for being authentic.

12: Learn to Have Fun

In the year 2019, Musk hosted the Bonus Meme segment of Meme Review, a popular series hosted by the most watched Youtuber PewDiePie who's real name was Felix Kjellberg. This was an additional major milestone for the site, considering that PewDiePie was trying to convince Musk to accept hosting the segment for a period of one month, believing that it would be the perfect complement to Ben Shapiro, the previous host. Ben Shapiro.

On the show Musk was a part of, Musk was joined by Justin Roiland, a co-creator of the popular TV series Rick and Morty. The two of them sat in a green space and discussed memes, as the name to the YouTube series suggests. They evaluated the memes using the scale of 1-10. The term meme refers to an inside joke that is posted on the internet, or an image that is easy to recognize even if it's been remixed. The idea behind memes is to provide an implicit joke triggered through visual stimulation, like clothing that has artwork by Daniel Johnston on it makes many Americans feel like Kurt Cobain. The memes play on the connections we make and instead of being used to foster an atmosphere of intimacy, or even to promote the work of an artist, memes are designed to entertain you while you are slamming your phone across your face while trying to fall asleep.

The majority of the memes were rather highbrow in terms of memes. One meme was a reference to the game of chess. The Queen of England and a bishop, stood in black squares in a room that had the floor covered with a checkerboard. The caption was "Be aware that she is able to change direction at any time." Musk loved the meme and gave it 9/10.

The most controversial meme in the review is one of the images depicting a deer drowned in a swimming pool. It includes a caption reading "Why my dolphin doesn't work lol". Initially, Musk was unsure of the meaning behind the image and he broke out laughing. Roiland was shocked however, he still lets out a few laughs before explained the Musk the image they're seeing. It takes a while to get Musk to realize that it is a real deer, but when Musk realizes what the image is, he chuckles even more and then breaks into the most genuine laugh he's ever experienced on the show. After Musk had calmed down and was able to look at the photo without chuckling at the absurdity in their job, they chose to not rate the meme.

Within the first day of that show's episode Meme Review being uploaded, the deer became a viral meme on the web. Musk's laughing towards the deer was absurd, and his continued to laugh was just as absurd. The comedian did not apologize for Musk's behavior that makes the clip of the show more entertaining genuine laughter was experienced by Musk in the meme. This is why the web was so fascinated by the meme and shared it to share the bizarre reaction.

It's even more make him more human Musk as it shows his natural humour. It was not an interview. It did not have anything to do with Tesla and it didn't advertise anything, it was Musk taking part in the excitement of younger generations. It was his evidence that he's not any superior to other people and that his humor can help contribute towards Meme Culture in as healthily the same way as everyone else.

For some overbearing officials as well as critics, the action appears to be a bad one from Musk since it shows an aspect of him that is not a positive one. Musk's wisdom lies in understanding that negative media is inevitable even when trying the best you can, there will be people who can find flaws in your work and persona.

The end result is that history will always side with the good humor. If you're taking yourself too seriously it is unlikely that you will get the respect of the public or many of your acquaintances. You must be able to relax and play with your emotions and show that you are also driven by joy and relaxation. By doing this, you demonstrate your values to others, and it demonstrates that you're an individual with solid substance if you are able to do things that humanity is doing. How can someone who wants to be a part of the growth of the internet, not have any curiosity about memes?

Musk chooses to be part of in on the excitement instead, and become component of technology Musk is planning to develop. There's no need to go to Mars when there's nobody to go there, and no reason to construct an internet when there's no user to make use of it. By putting himself in the shoes with the average user, Musk communicates his interest in the technology , far beyond the financial gain he'll earn by producing it. Musk demonstrates that he would like to make a difference in the lives of people because Musk will be a part of the peoples lives.

13. Let The Deer Smear Go The right people will laugh

After chuckling at the deer who drowned, Musk did not demand to cut his joke. Instead Musk let the video play, even though there was plenty of time to shoot again and edit it again. The video was released to the world wide web and the reaction of the actor was viewed as a viral event, and it was reported by media and shared worldwide by users.

There was no statement Musk made Musk following the incident for damage control, neither. The internet was able to do whatever they liked with Musk's image has only grown due to the willingness of him to participate his personal experience.

Do not put off doing something due to fear of the negative reactions. People who are right for you will be able to understand and connect with your experience. Don't be worried about how people perceive you.

13 Keep Things Small

One of the most crucial aspects of establishing and running successful businesses is managing expenses. Not just must this be accomplished for the resource management, but in teams. The strength of a company is only as the people that compose their workforce.

Musk created a business which sold for millions dollars. The team consisted comprised of six people with two of them being close contacts, three of them were salespeople. In Tesla Musk, Musk was quick to eliminate people who did not fit into the corporate culture.

There's no room for excess fat if you are looking to make making money. If you are prone to go, the higher you'll be. Make sure your connections are tight and don't share your ideas or resources. If you do this you'll save money as well as time as well as emotional stress.

Another advantage of having a smaller circle is that they are loyal. Over time, Thiel has been an essential ally to Musk despite the fact that they were initially rivals. The past experience and the familiarity of Thiel can be a huge factor when seeking out people you can count on in moments of emergency.

14 Retain Confidence

Musk was able to sell his ideas because the man believed in his own abilities. When he discussed his ideas with clients, there was no doubts about whether the goals he set out to achieve was possible to achieve. This gave Musk numerous opportunities throughout his professional life.

Do not let those you would like to collaborate with to believe that you'll be a failure. Be confident and show it. There is no need to worry about the errors if you are able to get the next day and keep committing to your objectives.

15: Do What You Have To

Musk's relationship with subsidy programs is a complex issue. Musk doesn't believe that subsidies are good for the economy or essential in a market-based market, but he utilizes government assistance as well as contracts with his business. We have ideas, but we don't depend on the realities to conform to our ideals.

To manage the companies Musk wanted to manage and achieve the goals he desired to accomplish Musk needed Musk to cooperate with the government, which Musk did. It would be incredible to see other space exploration companies which would be supportive of SpaceX however, there wasn't. If Musk wanted to develop space-related projects, Musk would have to cooperate with NASA. If he was planning to construct Model S, Model S, then he'd require a loan.

Reality is as it is and no one can manage it. It is more comfortable for us to be in a place that no one has stolen however in the event that you believed that there was no one stole, then you would not be able to reside in a city. Do your best however, don't sacrifice they if you're looking to make a difference in your life. It's better to think about how you can contribute to the progress of the things you wish to see rather than to be apathetic as if people don't want to be able to play the way you'd like them to.

All of us must become more mature and confront the fact that our thoughts aren't a direct reflection of the world that surrounds us. Our thoughts and beliefs have nothing to do with the world. Even the their expression can be considered an action. To confront reality, we need to overcome our egos and live by the rules it is governed according to, not how we wish it to operate. If we don't understand this, we'll be in danger of failing.

16: Better be an optimist and wrong rather than a pessimist or right

Musk believes it's best to be optimistic in dealing in the global arena. There are plenty of things to consider and there are numerous threats that humanity is likely to face however, it's best to tackle the problem with optimism. This outlook has been his hallmark throughout his professional career as well as throughout his life.

Musk has dreams and believes he will achieve his goals. When he faced the destruction of Tesla and his company, he didn't begin screaming in the public. Instead, he dealt with the pressure with grace and poise that only a only a few people could match. This enabled him to handle the crisis in a manner that was respectful and professional.

As well as overcoming difficulties, the goals Musk had set out to achieve were not considered feasible. He believed they were achievable and was willing to find ways to improve upon what was available and he succeeded. It requires a certain confidence to accomplish something that nobody else has done before. Even if we have to worry about the impact of fossil fuels but it's still an exciting effort to launch spacecraft. There are many things to think about but they should not hinder us from doing so.

17. Put your faith in your character

In his entire life, Musk has had an hostile behavior. He's argued with journalists and sued those who have criticized his work. If he's angry and irritated, he's likely to react and isn't shy about using harsh words. Musk doesn't often apologize but instead, he let the words he spoke out remain.

A lot of weaker people would give up under pressure from outside to alter their behavior, but Musk let his emotions remain as they are. If he says something, he intended to say it in a way that is meaningful to him. It shows a commitment to oneself and a determination that allows us to be awed by Musk.

You are who you say and the people respect them. If you obscure your actions through apology or compromises that you do not believe in the truth, people will not believe in your statements. Being consistent shows integrity and is highly valued. We like to get to know individuals for who they truly are.

18. Trust in the Law

Although some of the decisions made by the courts were questioned and questioned, such as the decision to side with Fisker in the case, Musk still remains positive in his assessment of his position on the American justice system. He says that judges are deeply committed to justice. They easily could have earned more in the field of law as attorneys.

Musk has never suffered any major legal challenge and has gotten through his way through American customer complaints. There were lawsuits filed against Tesla due to a myriad of incidents, including when someone driver of in a Tesla vehicle got sleepy at the wheel, and later hit and killed a cyclist. The driver then decided to file a lawsuit against Tesla and cited "the the new smell of cars" to explain the fact he was asleep when he struck the cyclist. A judge dismissed the case right away as absurd.

Through working within the systems and staying in the right direction that he was working in, Musk never had to face the prospect of being pursued by the law. His work is legitimate and he has shown respect when working with the authorities, so there's no reason why there shouldn't be any issues. Being honest is a smart choice because there's no reason to be afraid of showing your hand. If you're honest and transparent about your goals and plans and strategies, you'll live happy.

19. Be Unpretentious

It was easy to Musk to speak in a language nobody could comprehend. Musk learned how to program through books before at the age of 10 years old. Musk is a master of technology and physics that only a few could hope to attain. Instead of speaking at this scale, Musk speaks like anybody else. Musk curses, uses ordinary terms, and he expresses himself through references to culture that are real and humanizing.

Through his appearance in a way that is normal Musk performs himself huge favor. That's what's transformed him from a geek to hero. He was able to communicate extremely complex concepts in a manner that an audience could appreciate, and so his popularity increases.

This method for communication can be described as the most effective type of communication as it is widely distributed to the public. Since the internet has made media production and sharing more accessible and easy to do, public opinion masses is now a crucial factor. A variety of evidence points to the awareness of the power of social media as the reason that gave the presidency of both Obama as well as Trump their respective presidency.

Stay humble and you'll be a competent leader. People like to hear from those who can comprehend.

20. Build What You Would Like

Musk was not keen on dealing with the traffic, so he decided to start digging a hole. Musk was looking to pay online for people which is why he started X.com. Musk desired a car of luxury that was powered by the electric power of the motor and so the car was built by him. All the things Musk desired, he built himself by working hard.

They're just dreams until fulfilled. The lighter was a fantastic invention in the course of the past. The world can be better, all it takes is the effort to achieve it. If you're looking for it you're probably looking at something that a lot of people would like to have. If you are able to realize it and you can make it a reality, you will be able to be able to attract thousands of people to be part of your design.

Chapter 8: Little Known Facts

While attending University of Pennsylvania University of Pennsylvania, Elon Musk lived with Adeo Ressi. She became his closest friend at the time. In October 2001, when Musk was trying to acquire rockets made in Russia, Ressi accompanied him in the search. They've remained close and while at college, they operated an organization similar to a club by leasing an off-campus house. They hired bouncers and cleaners and successfully ran the club. The goal was to get to know people and get acquainted with campus policies However, they also managed to earn money for themselves.

During his time as the CEO of X.com there was a rumor that circulated around the company that Musk was so enthusiastic about using the term X.com was that he made a payment of $1 million to acquire the rights to the name. These rumors circulated through in the office just before the board decided to eliminate Musk and then changed the name of the entire business PayPal. The result was a stronger image for the brand, but it also has caused people to associate PayPal solely with the process of individual-to-person transfer.

Following the PayPal sale the entire web was set to be transformed. It is not widely understood how important this transaction changed the face of media and technology. When these investors got their share, they made investments into the media platforms users continue to utilize to support their own lives and communicate with others in their lives.

To begin, Yelp! was founded by Jeremey Stoppelmen and Russel Simmons. The company were also given $1million as an capital by Max Levchin, the CTO of Confinity. It was created around the idea of being able locate doctors via the internet through a referral service. Today, it has more than 100 million users, which is more than eBay was at the time PayPal was bought by eBay. It has reviews on various kinds of businesses in every American city as well as other cities around the globe. The reviews range from parks to restaurants, stadiums and even restaurants. In a way it's the complete vision that Musk was working towards when he created Global Link, which allowed interactive maps that showed companies. In 1980, such systems were impossible. Nowadays, consumers are more likely to steer clear of businesses with poor reviews.

Peter Thiel used his profits from the sale of $55 million, to launch Clarium Capital, an investment hedge fund for management. Additionally the fund, he also founded Palantir Technologies, software and services firm that is focused on Big Data Analytics. The most profitable investment opportunities for Clarium Capital was being an early investor in the just-starting company, Facebook. This web site, Facebook, has grown to become one of the top websites on the planet. It boasts 2.32 billion active monthly users, and 1.15 billion per day users on mobile devices alone. One fifth of the traffic to websites in America is done on Facebook. About 40% of businesses depend on Facebook to spread their message. Thiel's investments and businesses have allowed Thiel to own his net worth to exceed $2 billion in 2018.

Linkedin was founded by Reid Hoffman, the executive vice president of PayPal at the moment of PayPal's sale to eBay. There are more than 154 million accounts available on the site , primarily from the American employees in the United States alone. 3 million jobs are posted through the platform every month. 50% of college educated Americans already have an account. Hoffman is the 37th most visited website in the world. Hoffman is worth more than $3 billion.

Luke Nosek, Peter Thiel Peter Thiel, Luke Nosek as well as Ken Howery went on to co-found The Founders Fund, which is mostly fictitious due to its investment in SpaceX. The venture capital company is also investing in Spotify as one of the most well-known streaming platforms in the 2010s. Airbnb basically the sole popular app for hospitality and Lyft the most popular ride-sharing service, which is one of the few legitimate options of payment Uber offers. This fund, known as the Founders Fund has built some of the most popular utility apps for smartphones, greatly increasing the capabilities of smartphones.

The sale of PayPal affected the course of the internet by giving Chad Hurley, Steve Chen and Jawed Karim the money to build the second-most well-known site on the planet, YouTube. The only website that is more popular than the video sharing site is operated by the same company that bought YouTube at the end of 2006, Google. Every minute, YouTube users upload over 400 hours of video to YouTube in the course of a day, YouTube users view more than a billion hours on it. There's been no other site like it, and YouTube is said to be the most influential website online.

Alongside being the beginning of Musk's capacity to pursue his own interests and interests, the acquisition of PayPal has had a massive impact on the growth of internet technology. After enduring the experiences of building a billion-dollar online business employees quit PayPal with the ability to come up with whatever ideas they could come up with. The ideas they had sparked into the utility that allow independent creators to earn a living, professionals get jobs, children are raised and commuters are able to find transportation in cities to go about their daily lives. The internet as well as the world at large is not the same following the impact on this PayPal sale.

The first car to go into space was launched with the help of Elon Musk in the year 2018. Elon Musk used the Tesla Roadster as the dummy payload to test the flight by the Falcon Heavy. The driver's seat is Starman who is a spacesuit-clad mannequin. The Roadster was the car Musk was using on his way to and from work. Some criticize this choice because it could be viewed as an element of debris from space, but as Musk stated in 2008, there's such a vast space and such a small amount of debris, that there's an extremely low chance of spotting problems when launching a rocket and craft that transport living cargo come with meteorite shields to be used to shield against impacts. Musk's explanation, however, is not a defense rather a statement that he is simply trying to rekindle curiosity in creation of space-related technologies. Musk says that we'll reach Mars.

Musk believes that life is an illusion. According to him that argument is strong due to technological advancements suggest that there will come an point at which video games will not be able to differentiate from reality, or civilization will end. Because civilization isn't eroding, and our world appears to be functioning, Musk believes that, most likely, we're in a simulation since we are real. The possibility is that this is an illusion, and that we are moving towards either the ultimate game technology or the end of the world.

But, it is higher that this is a simulation. He believes that the reality itself is a boring experience with next to no activity with it. Reality is the summation of the finest aspects of reality. Simulations are designed to be more entertaining than the reality they are derived from, so our reality, even if it's a simulation was designed to be enjoyable.

At numerous instances, Musk has been found using the term America as the greatest nation that has ever existed. Musk believes that, without that country, democracy could have been lost across the world and, if not during after the end of World War I, then after World War II, if not for and after the Cold War. According to him it was America which has preserved people's voice masses and also the position of citizens in the contemporary world.

It is interesting to note that Musk declares himself socialist. Musk differentiates his version of socialism by saying that he's "not one who shifts resources from the most productive to the least productive, while pretending to be doing good but actually doing harm" instead "true socialism is a search for the best common good for all." While this may be seen as being in opposition to the American ethics, it's likely a reaction to the advancement autonomous vehicles. The greater success Tesla is able to achieve with its self-driving cars, the more jobs that will be lost by automated vehicles. If Tesla trucks take over truck drivers, there will be a major economic disruption that will result in the displacement of thousands. Perhaps, Musk's inclination towards UBI along with the social labelists may be a response to this fact and he's trying to for ensuring that those who are impacted by his success will have a safety net to be able to fall into. In any case, the technology will develop and their business will be at risk. It's best to prepare to deal with the situation.

However, Musk donates to both the Republican and Democratic party in America However, he tends to give more to the Democrats. Musk explains that in order to speak out in America it is necessary to give money and that's a sad fact. To make a difference it is necessary to participate in the finance game otherwise, the system that runs the country will not be able to take your opinion into consideration. He is a person who believes he's in between the two political parties, politically liberal, but fiscally conservative. His priorities don't align with political boundaries, but they encompass the entirety of America.

In the past, Musk served as an advisor to the president Donald Trump. He was not a fan of the president , but said that it was essential for the president Trump to be with as many rational voices as is possible. Musk has resigned from the post after the Trump administration pulled from the Paris agreement. Musk wishes more politicians were educated in science. He is awed by China for the amount of political leaders who have degrees in science.

Musk believes that humans are already cyberborgs. He views smartphones to extend self, and sees smartphones as extensions of consciousness. This is a controversial viewpoint, but it has merit. Smartphones are the cause of much of the burden that the imagination and memory were previously.

AI is a major concern for Musk. He has tried for years to stop the growth of AI, claiming it to be risky once it reaches the point of being complex. He fears that humanity will develop something that is impossible to manage or control in the end, and ultimately become the entire world.

In during the Obama Administration, Musk met with the White House to discuss AI and his fears regarding the future. Musk attempted to explain AI was among the most significant risks humans could face if it were allowed to evolve as it did.

The world has not did listen to Musk's warnings, however. AI continues to grow at alarming speeds and the civilization could be in danger of extinction. In the most practical sense, a lot of occupations are in danger. AIs have the ability to be programmed to learn the art of welding. AIs can drive the car, and there are AIs that cook. The majority of physical or mechanical tasks can be accomplished more efficiently by machines.

Machines are more accurate in repetitive tasks. They are able to work for a long time and don't have to be compensated. There will be a long time before AI is competitively viable in the market. And when it has a complete grasp of work that is being done, the world will appear completely different. A number of candidates advocate for UBI which stands for Universal Basic Income, in anticipation of this upcoming change. Musk is a strong supporter of this plan, but it did not work when test-driven in Scandinavia.

Musk has the Jaguar Series 1'67 E-type Roadster that is a totally manual gasoline car. Musk refers to the car as being his initial obsession with automobiles. Inside the car, there are no electronic devices. There is no electronic system. McLaren F1 was crashed in 2000, when Musk did a flip airborne and caused it to spin until it was landed in the same direction that he was driving. The crash happened because Peter Thiel asked, "So what do we do about this." Musk replied with the famous final words, "Watch this."

He caused the car to spin forward, and then changed lanes on Sand Hill, a famous road in Silicon Valley, but ran into an embankment that was 45 degrees and that's what caused the crash. After the crash the vehicle was destroyed. The two of them Thiel and Musk were able to recover.

When Musk was a child, about 5 or 6 years old, Musk realized that he was different from those who were around him. His brain is hyperactive which means that there's always a flow of information that flows throughout his mind. Ideas come to him constantly and he realized that others weren't experiencing similar experiences. When he was first coming to an understanding of what was happening and tried to conceal his thoughts from the world to conceal his intellect because he believed that he was insane and could be thrown out.

Chapter 9: Entry Into The U.S.

After 2 years of Queen's University, Musk moved to the University of Pennsylvania. He was a major in two fields and yet there was not a lack of the balance of work and life. With an understudy of his own and a 10-room clubhouse, which was employed as a specially-appointed nightclub.9

Musk was awarded a four-year certification in scientific studies in physical science along with an education of four years in liberal arts with financial aspects at the Wharton School.10 Both majors were a precursor to Musk's future career path, however the physical science was the most significant influence.

"(Material scientific research is) an excellent system of reasoning," he would agree later. "Reduce the things down to the essential elements of understanding and think from there. "11

Eminent Accomplishments

Musk was just 24 years old age when he relocated to California to pursue the Ph.D. on applied physical sciences at Stanford University. As it happens that, with the Internet exploding in the background and Silicon Valley blasting, Musk was a pioneer through his head. He quit at the end of his Ph.D. program after only two days.12

X.com

In 1995, having $28,000 and his younger brother Kimbal close by, Musk began Zip2, an online programming company that assisted newspapers in making online cities guides.13 It was in the year 1999 that Zip2 got acquired by Compaq Computer Corporation. to run its AltaVista index of websites with a purchase price of 341 million.14 Musk utilized his Zip2 purchase to fund X.com which was a fintech venture prior to the time that the term was widely used.

X.com joined forces with a cash transfer company called Confinity which later became Confinity, and the resulting firm was later referred to by the name of PayPal.15 Peter Thiel removed Musk as PayPal's CEO prior to when eBay (EBAY) bought the installments company for $1.5 billion. However, Musk was still a beneficiary of the buyout via his 11.7 percent PayPal stake.1617 "My earnings from PayPal after the charge were around 180 million dollars," Musk said in an annual meeting in 2018. "$100 million of this went into SpaceX and $70 million went into Tesla as well as $10 million to SolarCity. Additionally, I in the real sense had to obtain cash to lease."

TESLA

Musk was involved in the electric vehicles venture as a financial supporter in 2004, eventually contributing around $70 million before becoming a partner in the engineers Martin Eberhard and Marc Tarpenning to aid in the running of an organization that was then called Tesla Motors. Following the demotion of Eberhard in 2007 due to an escalating conflict, Musk took command as the CEO and as an item engineer. Under his direction, Tesla has turned into the most important automaker in the world and also one that has taken over its most renowned brands.20

Along with delivering electronic vehicles Tesla is able to maintain a prominent presence sun-based energy sector, because of its security the rights to SolarCity.21 The company is currently working on two solar-powered batteries powered by batteries. The smaller Powerwall was developed to power homes and off-the-matrix usage, whereas the larger Powerpack is designed for commercial or utility use.

SPACEX

Musk made use of the bulk of the earnings of the sale of his PayPal stake to create Space Exploration Technologies Corporation, the engineer of rockets, commonly referred to as SpaceX. According to his own records, Musk burned through $100 million to start SpaceX in 2002.

Under Musk's direction, SpaceX handled a few notable agreements with U.S. Public Aeronautics and Space Administration (NASA) and the U.S. Space flying corps to design space-based rockets. Musk has revealed that he will send an space probe to Mars by 2025 , in an effort in cooperation with NASA.24

TWITTER

A constant banner of the organization that informs, Musk revealed a 9.2 percent ownership stake of Twitter at the beginning of April in 2022. Twitter responded by providing Musk an appointment on its board, which Musk accepted prior to a decline in days following the fact. Musk then wrote an "giant squeeze" letter to the board of Twitter suggesting that the company be bought for $54.20 per share. That's an 18% increase over its market value, however it would be a 22 percent discount on Twitter's share of the value one year earlier.

The Twitter board adopted an arrangement for death to prevent Musk from acquiring a greater stake. The board at any rate, Musk's idea after he disclosed $46.5 billion of pledged support for the agreement in an protections filing.2526

Individual Eccentricities

On September. 7 7, 2018 Musk has been smoking marijuana in the shot-making session to record podcast. podcast.27

Just a month earlier, Musk posted a notorious tweet that claimed he was contemplating about going public with Tesla public and that he had received the necessary subsidization. Musk consequently settled an U.S. Protections and Exchange Commission (SEC) protest, claiming that he deliberately misled financial backers through the tweet. He settled for an amount of $20 million as the same punishment for Tesla and apprehension to let Tesla's lawyers approve tweets that contain corporate information prior to publishing.

In March 2022, Musk made a court motion to stop the assent order in the case.28 In April 2022, Musk made reference to SEC controllers who were working on it to be "rats."

DIFFERENT activities
Hyperloop
Essential Articles: Hyperloop and Hyperloop case competition

In 2013, Musk revealed plans to develop a variation of the Vacutrain (or the vacuum tube train) and appointing twelve designers from Tesla as well as SpaceX to design the decent infrastructure and create the initial plans. The 12th of August 13th, 2013 Musk disclosed the idea of the Hyperloop. The initial plan for the Hyperloop was released in a whitepaper published via SpaceX and Tesla's Tesla and SpaceX journals online. The archive looked into the concept and provided a conceptual plan in which a similar vehicle framework could be operated across both the Greater Los Angeles Area and the San Francisco Bay Area at the estimated cost in the range of 6 billion dollars. If it is technically feasible at the prices the author has cited will allow Hyperloop journeys less expensive than any other mode of transportation that covers such vast distances.

In June of 2015, Musk reported a plan competition for students and other understudies to build Hyperloop units for the SpaceX-supported mile-long track for an ongoing 2015-2017 Hyperloop case-based competition. The track was put to use in January of 2017, and Musk also reported that the company had started an effort to make a passage with Hawthorne airport as its goal. The track was completed in July 2017. Musk stated that he obtained "verbal approval from the government" for the construction of the hyperloop that would run that would run from New York City to Washington, D.C. This would stop the train at both Philadelphia as well as Baltimore.The track was operational in the spring of 2017. A mention of the project to build part of the DC to Baltimore portion was removed from the Boring Company website later in 2021.

OpenAI

As of the end of 2015 Musk announced the creation of OpenAI which is a non-profit AI research institute that aims to create counterfeit general knowledge that is expected to be secure and beneficial to the human race. One particular focal aspect of the organization is to "balance large corporations [and even government agencies] who could gain lots of power by using brilliant frameworks". In 2018, Musk quit the OpenAI board to steer clear from potential future conflicts in his role as the CEO of Tesla as the company progressively became involved with AI via Tesla Autopilot.

Tham Luang cave salvage and maligning case

In July of 2018, Musk set up for his team to create an unassuming salvage unit to assist in the recovery of kids trapped by an overcrowded Sinkhole that was discovered in Thailand. Richard Stanton, head of the international salvage jumping organization requested Musk to assist in the creation of a smaller sub to be used as a backup in the chance that flooding was to worsen. It was named "Wild Boar" after the children's soccer team it was designed to be five feet (1.5 meters)12 inches (30 cm) wide fixed tube that weighed about 90lbs (41 kg) driven by jumpers to the back and front with sections to allow loaders on the jumpers, aiming to address the problem of safely removing the children. Engineers from SpaceX as well as The Boring Company assembled the tiny submarine from an Falcon 9 fluid oxygen move tube in just eight hours and then transported the submarine to Thailand. At this time, in all likelihood, eight of the 12 children were proactively protected using masks that covered their faces as well as oxygen with sedation therefore, Thai experts decided not to use the submarine. Elon Musk was later among of the 187 who were awarded different distinctions to the King of Thailand in March 2019 for his assistance in the rescue effort such as that of the Order of the Direkgunabhorn.

Vernon Unsworth, a British caver in the sport who has been studying the cavern over the last six years and played a crucial role in the salvage, was rebuked by his submarine's appearance on CNN in the hopes that it could not be more than an advertisement with no chance of coming out on the top. He also said that Musk "had no idea the cavern's entry as" and "can put his submarine wherever it is hurting". Musk claimed via Twitter that the gadget could have worked , and referred to Unsworth as a "pedo person". The tweets were then deleted along with a tweet, in which he advised another pundit about his gadget "Remain in tune jackass. "[200On the 16th of July, Unsworth expressed that he was considering legitimate activities.

Two days later, Musk posted a resolute opinion regarding his comments. On the 28th of August in 2018, based on the analysis of the author of Twitter, Musk tweeted, "You do not think it's odd that to not have sued me?" A few days later, a letter from the 6th of August by L. Lin Wood, the lawyer of the hero, surfaced which showed that he'd been working on preparing for a slander lawsuit.

In the midst of this James Howard-Higgins had a message to Musk declaring that he was an examiner for private parties and made an offer to "dig deeply" into the past of Unsworth, which Musk acknowledged. Higgins was later found to be a criminal indicted with a variety of fraud charges.[] In August, using techniques he created during the investigation, Musk sent a BuzzFeed News correspondent who had elaborated on the subject an email with the subject line "in private" and a message telling the reporter to "quit protecting kids from attackers, you're flogging a poop hole" and that it was an "solitary white male from England who has been traveling into or living in Thailand for between 30 and 40 years to ensure the safety of Unsworth... up to the time he moved into Chiang Rai for a youngster woman who was about 12 years old at the time. of age at the time." The 5th of September, the journalist tweeted a screen shot of the email, stating that "In private, there is a two-way agreement" that Musk "didn't agree to."

On September 1, Unsworth recorded a defamation lawsuit in Los Angeles government court. With all respect, Musk contended that in shoptalk usage "'pedo fellow is a common affront that was used within South Africa when I was growing in my teens... it is inseparable from 'dreadful elderly and used to criticize the appearance and manner of a person. "[216The criticism case began in December of this year with Unsworth seeking $190 million for damages. In the initial hearing, Musk apology to Unsworth yet again for the tweet. On the 6th of December the jury decided in favor of Musk and concluded that the jury that he did not have any responsibility.

2018 Joe Rogan digital broadcast appearance

On the 6th of September the 6th of September, 2018 Musk was on The Joe Rogan Experience digital show and spoke about various topics for over two hours. At the time of the interview, Musk inspected a solitary puff of a stogie that was comprised, Joe Rogan asserted, of weed and tobacco. Tesla stock fell after the incident, which was in accordance with the assertion of the departure of Tesla's VP of general cash prior to the day. [220][221The Verge reported that Fortune was pondering whether the pot's use might impact SpaceX agreements in partnership with The United States Air Force, however, an Air Force representative let The Verge know that there was not an examination or investigation. The Air Force was all the when handling the issue. In an hour-long interview with The Verge, Musk stated of the incident: "I don't smoke pot. It should be evident, I don't have a idea how to smoke marijuana. "[224][225]

Music adventures

On March 30th 2019 Musk released a rap song, "Tear Harambe", on SoundCloud as Emo G Records. The track is a reference to the death of Harambe the gorilla at the Cincinnati Zoo, and the subsequent "boring" Internet melodrama encompassing the event it is performed by Yung Jake, written with Yung Jake as well as Caroline Polachek. The track was composed by BloodPop. The 30th of January in the 30th of January, Musk delivered an EDM track, "Don't Doubt Ur Vibe" featuring his own lyrics and vocals. [228] Although The Guardian pundit Alexi Petridis described it as "unclear... of umpteen pieces of uninteresting, but well-equipped bits of room electronica uploaded elsewhere on Soundcloud",[229[229] TechCrunch stated that it was "not an accurate portrayal of the genre".

Non-benefits and gifts

In 2012, Musk signed the Giving Pledge and subsequently vowed to dedicate the bulk of his wealth to worthy projects during his lifetime or through his will. In 2014, Musk donated 1 million of his wealth to an art museum dedicated to Nikola Tesla. In 2020, Forbes awarded Musk an overall score of 1 since Musk had given less than 1 percent in his total net worth. Then, in November of 2021 Musk donated $5.7 billion in Tesla's earnings to charities. He has also enhanced prizes offered by the X Prize Foundation, remembering $15 million to stimulate innovation to address ignorance, while also donating $100 million as a way to pay for the carbon capture technology that he has further developed. Following the 2022 Russian invasion into Ukraine, Musk worked with the sending of Starlink frameworks purchased from other European countries and private finance to Ukraine to allow access to the internet and communications within the harmed country. [238238 Ukrainian President Volodymyr Zelenskyy said thanks to Musk in the past for his assistance to Ukraine and proclaimed further talks between the two countries on space projects to be held following the conflict. [239][240]

Musk Foundation

Musk is the founder in Musk is the founder of Musk Foundation,[241], which states its mission to provide solar-powered structures for energy in disaster zone as well as aid research, development and support (for these areas as human space exploration as well as the study of pediatrics, sustainable power, and "safe human-made reasoning") and also research and development of instructive goals. In the past two years, the foundation has made over 350 pledges. Nearly half of these were for research or instruction non-profits. The most notable recipients are Wikimedia Foundation, the Wikimedia Foundation as well as his school of graduation , the University of Pennsylvania, and Kimbal's Big Green.] Vox described the institution in terms of "practically entertaining in its simplicity however, it is a bit cloudy" noting that the site contained only 33 words in plain text. [244] The organization has been criticized for its relatively small amount it that it gave away. From 2002 until the year 2018, it distributed $25 million to people who were not benefiting nearly 50% of which was given to Musk's OpenAI which was then an association that was not a benefit.

TWITTER
Musk is a user of the entertainment platform online Twitter with greater than 80 million fans. Musk created the primary tweet of his account in June of 2010. [248] He shares pictures, discusses financial affairs, and sometimes makes remarks about current social and political issues.
Musk has been the subject of some debate because of his use for the stage. In August 2018, Musk claimed in his tweet that he

would be taking Tesla public for $420 per share, which was a joke about pot. A SEC investigation concluded that the tweets did not have a base in fact and caused harm to investors who were financial backers, leading to small fines for Musk as well as Tesla in the amount of 20 million dollars each. The agreement also contained the provision that Musk could legally to endorse tweets concerning Tesla before the deadline. In 2022, Musk was accused of a lawsuit by Tesla investors for the tweet. Musk continues to insist his claim that it was "worth the price". [251] In 2020 , a tweet from Musk declaring that "Tesla stock costs are excessively high" reduced the value of the company by 14 billion. Musk had more issues 2021 when he posted an opinion poll on whether or not to sell 10% of his shares within the company before deciding to do so. This led to an SEC insider exchanging investigation of Musk and his brother Kimbal which was based on the possibility that Musk had told his brother prior to the time he would tweet the survey. Another sign of discussion include his tweets regarding digital money like Dogecoin and Ethereum and have caused their worth fluctuate, [252][253]as in tweets that minimize the severity of COVID-19 and analyzing lockdowns, like one that contrasts Canadian chief of government Justin Trudeau with Hitler over the immunization requirements. [251]

Musk tweeted as early as 2017 to announce his interest in purchasing the platform. In January 2022 Musk began buying significant amounts of deals in the company, and he reached 5 percent stake in the company on March 14th 2022. [254254 Musk was able to reach a figure of 73,115,038 offers as of April 1st 9.13 percent of Twitter's overall offerswere valued at the time as $2.64 billion, which made Musk the largest investor in the company. The report claims that Musk did not record the essential administrative tasks to inform the SEC in not less than 10 days from the time his share of the business outperformed that of 5%, which is an breach of US protections laws. When Musk publicly announced his desire to participate in the Securities and Exchange Commission 13G documenting on April 4, 2022 Twitter shares suffered the biggest intraday rout since the company's IPO in 2013, when an exchanging flurry resulted in the price of the offer increasing by as much as 27 percent. The revelation that Musk was acquiring a massive stake in Twitter was prompted by Musk's tweets of the 25th and 26th of March, in which he questioned Twitter's commitment to allow speech and claimed that he had been thinking of creating a virtual entertainment site for opponents. website, even though the comments were made following the fact that he had previously purchased 7.5 percent of the company.

In April, Musk agreed to an agreement which would have him appointed to Twitter's top management team and prevent him from taking more than 14.9 percent of the company's stock,[261][262but Musk decided not take on the role of board member until the agreement became enforceable the 9th of April. On April 13 Musk announced an offer of $43 billion to acquire Twitter by submitting an offer to buy 100 % of Twitter's stock with a price of $54.20 for each share. In a letter addressed to the board of Twitter, Musk expressed his desire to take Twitter to the next level: "[Twitter] will neither prosper nor fulfill its freedom of speech] culture in its current structure. Twitter ought to be changed into an entity that is privately owned. "[264Accordingly Twitter's board has adopted an investor rights plan to make it more expensive for a single financial backer to own more than 15 percent of the company without the endorsement by the board. In an TED appearance, Musk showed little interest in fighting web control around the world and said that "Twitter must comply with the laws of the nation". In the end Musk's fear of free speech is centered with Twitter's control methods.

On the 20th of April, Musk received subsidies that totaled $46.5 billion. The subsidizing accounted for $12.5 billion in advances to Musk's share in Tesla and $21 billion worth of value-based funding, such as the proceeds from trading Tesla shares.[] The 25th of April reported as if Twitter was prepared to accept Musk's offer. The following the same day Elon Musk was able to close his deal to purchase Twitter and make the company into private hands for $44 billion. In a statement, Musk said:[278][279]

Free debate is the basis of a majority rule system. Twitter is the digital town square in which issues crucial to humankind's future are discussed. I must improve Twitter more than at any other time in recent history by updating the application with fresh elements and making the calculations open source , increasing trust, eliminating spam bots and validating everyone. Twitter is a huge opportunityI am looking forward to working with the organisation as well as the local community of users to make it available.

The value of Tesla's financial exchanges fell by $125 billion in the subsequent day as a result of the agreement, resulting in Musk losing around thirty billion in his wealth.Then he tweeted an the analysis on Twitter chief Vijaya Gadde's messages at his more than 86 million followers that resulted in some of them to take part in bigoted and misogynist bullying against her.

Abundance

Musk has earned $175.8 million at the time that PayPal was sold the eBay company in 2002. The first time he was listed as a member of Forbes' Forbes Billionaires List in 2012 and had worth of $2.5 billion. [286]

In the first quarter in 2020 Musk was accounted for $27 billion in assets. [287] By year's close, his assets had grown by $150 billion, which is usually measured by his share of 20 percent of Tesla shares. In the course of this period, Musk's total assets were frequently volatile. In one instance, it fell $16.3 billion during September. This was the highest single-day drop throughout the history of Bloomberg Billionaires Index. In November of the month, Musk beat Facebook fellow patron Mark Zuckerberg to turn into the third most extravagant person in the world. After seven days, he beat Microsoft principal supporters Bill Gates to turn into the second-richest. In the month of the month of January in 2021 Musk had a net wealth of $185 billion beat Amazon planner Jeff Bezos to turn into the most lavish person around the globe. [291291 Bezos returned to the top position during the month that followed. 292] On September 27 2021 Forbes stated that Musk had assets worth more than $200 billion. Musk was the richest person on earth, following the Tesla stock soared. In the month of December 2021 Musk became the most prominent person with total worth of $300 billion. [294]

Around 3/4 of Musk's wealth comes his wealth from Tesla. [290The reason for this is that Musk isn't compensated from Tesla and he voted with the board in 2017 to agree on a pay arrangement with the board which linked his earnings to Tesla's revenue and valuation. The agreement stipulated that Musk could be eligible for the compensation in the event that Tesla is able to meet certain market prices. [295This was the largest ever arrangement executed between a CEO or board. [296] As part of the principal grant, which was announced at the end of May, 2020 was able to purchase 1.69 million TSLA shares (around one percent of the business) at a price below market prices that was approximately 800 million dollars. [296][295]

Musk has paid $455m for charges on $1.52 billion in payments between 2014 and 2018. In the report by ProPublica, Musk paid no personal expenses for the government in 2018. His 2021 duty bill was estimated at $12 billion, in light of Musk's offer of 14 billion in Tesla stock. [297]

Musk has many times been portrayed as "cash poor",[299][300 and has "pronounced that he doesn't care at any of the physical aspects of his wealth". The year 2012 was the time Musk signed The Giving Pledge and in May of 2020, declared that he would "sell virtually all possessions that are physical". [300][301] By 2021 Musk safeguarded his wealth by saying that he is "aggregating assets to help in the multiplication of life and expand the scope of awareness up to space". In the early decade of the 2000s Musk used to be an independent pilot, his primary aircraft at the time at that time was his L-39 Albatros, however he decided to stop directing in the year 2008. He has an aircraft that is owned by his personal luxury company, SpaceX. by SpaceX[305][306] that he obtained an additional flight during August of 2020. [307] The stream's heavy use of non-renewable energy sources The stream was north of 150,000 miles in the year 2018and has received an analysis.

Sees

U.S. governmental issues

Musk together with US Vice-President Mike Pence in 2020 at the Kennedy Space Center presently before SpaceX's Crew Dragon Demo-2 send off

In 2015, Musk stated that the fact that he is an "huge (however not top-tier) supporter of Democrats" however Musk also gives heavily to Republicans. Musk stated that monetary political commitments are essential to be able to speak out in Congress in United States government. 309 Musk has criticized Donald Trump for his position on climate change[310] and following his joining of Trump's two business advisory councils[311][312312 Musk quit both of them in June 2017 to contesting Trump's decision to exile from the United States from the Paris Agreement. In the 2020 Democratic primaries, Musk welcomed newcomer Andrew Yang and communicated help in his proposal for a widespread basic income. He also endorsed Kanye West's freedom mission during his general election. [315315 Musk has said his belief that a hypothetical government on Mars should be an immediate democratic. Concerning the Democratic proposal to increase charges on wealthy individuals Musk's reaction has included basic statements about strategy and attacks on supporters like Ron Wyden, Senator Ron Wyden. [317][318]

Then, in November of 2021 Musk has been sacked following a scathing attack on U.S. Representative Bernie Sanders on Twitter. Sanders posted a tweet on Twitter which read "We should demand that the extraordinarily wealthy receive their fair share. Period." Musk then at that point said: "I continue failing to be aware the fact that you're alive."[319][320][321]

Texas"social strategies

As of September 20, 2021 after the acclamation of Texas the strict early termination restrictions, Texas Governor Greg Abbott declared his belief that Musk and SpaceX stood by Texas' "social approaches". As such, Musk expressed, "as as a rule I think that the government should not dictate its decisions to individualsand, when doing so, it should try to increase their joy at being together. With that said I'd like to stay out of politics."

Finance

Unregulated economy

Musk has stated that he doesn't believe authorities in the U.S. government ought to provide endowments to companies and instead use a carbon tax to stop bad behaviour. [382][383He says that Musk declares that an non-regulated economy will result in the best result and that creating cars that are earth-disappointing should have its own consequences. His argument is regarded as dishonest because his companies have received billions of dollars in subsidy. Additionally, Tesla made enormous totals from frameworks created by the government that have zero outflow credits, which were announced at California and at the United States administrative level, that enabled superior introductory buyer acceptance of Tesla automobiles, because the tax cuts provided by legislators allowed Tesla's electric vehicle to competitive, and in a direct correlation with gas-powered motor vehicles. Also, Tesla produces a sizeable part of its earnings from carbon credits granted to the company, through as well as the European Union Emissions Trading System as well as the Chinese public carbon exchange scheme. [388][389][390][391]

Assessment of tax

In a meeting in December 2021 with Christian moderate site The Babylon Bee, Musk complained the fact that it had become "progressively difficult to complete things" at the time in California. Musk was able to achieve his goal that he sell 10% stake of Tesla (then the largest vehicle company in the world) as well as some of it to pay back costs, and then moved his personal and Tesla's expenses house in California and moved to Texas to avoid the state personal assessment. Musk said that "California was once the location where you could find new opportunities, but now it's... changing to be more as a place in which there is a bit of excessive regulation, litigation, and overtaxation. "[392 In the same meeting, he stated, "At its heart progressiveness is disruptive, exclusionary and a source of contempt. It gives people who are mean... an opportunity to be vile and savage that is protected in a false sense of virtue. "[393]

Exchange

Musk is a long-standing opponent of short-selling He has repeatedly dismissed the program and claimed it should be illegal. Musk's opposition to short-selling is believed to be rooted in the way short-sellers often put together and disseminate untrue information about the companies which they believe are and are now undervalued. In the mid-year of 2021, he facilitated to allow the GameStop Short squeeze. [397][398398 Musk has also been consistently advancing digital currencies in his statements, affirming that he favors their legitimacy over conventional, recognized fiat currencies. Because of the erratic consequences that his tweets regarding them can have, his claims regarding cryptographic currencies are viewed as market control by pundits such as Nouriel Roubini. [401]

Innovation

Man-made consciousness

Musk has often spoken of the risks that could be posed by AI repeatedly, saying that it is the biggest threat for humanity. Musk's views on AI have provoked debate and been criticized by experts such as Yann LeCun. As reported in CNBC, Musk is "not typically regarded as a good choice" among members of the AI researcher community. [407407 Mark Zuckerberg has conflicted with Musk regarding the subject in the past, with Zuckerberg saying that Musk's advice is to be "pretty reckless". Musk's claims that people are living in a programmmatic experience are also being dismissed. [411][412]

Metaverse

The year 2021 was the month that Musk began his career. Musk was questioned about his view on the computer-generated metaverse driven by experience, Musk said that he "couldn't imagine any convincing metaverse scenario" and then added "I believe that we're a long distance from being a part of the metaverse. It's a fashionable. ... Yes, you can wear a television on your nose. I'm not sure what is what makes you 'in"the Metaverse". ... I haven't observed anyone strapping the mother truckin screen to their face all day without having any desire to go away. It appears that there is no way. "[413][414]

Global struggle

Bolivian rebels

On July 20, 2020 following to being requested via Twitter by one of the clients "it was not to the benefit of the individual" to allow Musk to "US government to plan an uproar over Evo Morales" in order for Musk to "acquire lithium" from Bolivia The money manager replied: "We will overthrow whoever we require! Take care of it." The tweet provoked debate and was later removed. [415][416]

2022 Russian intrusion into Ukraine

Musk condemned his 2022 Russian attack on Ukraine and announced measures to aid Ukraine's security by, for instance, providing Ukraine with gratuitous Starlink access.

Ukrainian the President Volodymyr Zelenskyy, by and by, expressed gratitude to Musk as he announced further talks between both. [417][418]

Populace

Musk has expressed concerns regarding the declining human population,[419][420] declaring "Mars is a place with no people. We want to see a lot of people to become the multiplanet society. "[421speaking at the Wall Street Journal's Chief Executive Council session in the month of December, 2021 Musk declared that the decline in population and birth rates is the most significant risk to the human race. [422]

LIFE INDIVIDUAL

Musk found his memorable wife, Canadian creator Justine Wilson who was attending Queen's University, and they got married in the year 2000. He was diagnosed with an illness called jungle fever during traveling for a long time in South Africa, and almost died. In 2002, their most loved child, Nevada Alexander Musk, passed in the midst of an abrupt infant demise (SIDS) at the age of 10 years old. [425] Following his death the couple decided to go through IVF to carry on with their family. twins Xavier as well as Griffin were born to the world in April 2004 and were followed by the trio of Kai, Saxon, and Damian in the year 2006. The couple split in 2008 and now share the responsibility for their five sons. [423][427][428]

In 2008 Musk started dating English entertainer Talulah Riley. The couple got married in September of 2010 at Dornoch Cathedral in Scotland. In 2012, Musk decided to divorce from Riley. In 2013, Musk and Riley were married again. The couple divorced in December 2014. he filed a petition to briefly to separate from Riley but in the end the matter was dismissed. The separation was concluded in the year 2016. [436436 Musk later was, at the time, began dating Amber Heard for a long period of time in 2017. He was reportedly chasing Amber Heard since the year 2012. [438Then, in 2017, Musk was later accused for the affair by Johnny Depp for engaging in extramarital affairs with Heard even though she was not yet unmarried to Depp. [439][440][441The couple Musk as well as Heard both denied having an affair. [442]

In May of 2018, Musk and Canadian artist Grimes revealed they were in a relationship. [443][444][445The couple Grimes had a baby on May 20, 2020. According to Musk and Grimes his name was "X A-12" However it would violate California guidelines since it had characters that weren't part of the modern English alphabet,[448][449[449] and was later changed from "X A-Xii AE". The name caused more confusion since AE is not an alphabet in the most modern English alphabet. The boy was identified as "X AXII" Musk, with "X" as the initial name "AE AXII" as the center name as a middle name, as well as "Musk" as the surname. [451(451) Musk confirmed reports that the couple was "semi-isolated" at the time of their September 2021 meeting. In a conversation in a meeting with Time at the end of December in 2021 Musk stated that he was not married. In March 2022 Grimes spoke of her romance and Musk: "I would presumably be referring to Musk as my sweetheart but we're very liquid." The actress also revealed that their most beloved daughter, Exa Dark Siderael Musk known as Y and born in the world in the month of December 2021 by a surrogate. In the following year, Grimes announced saying that she and Musk had split again "yet Musk is my best friend and has the most affection for my life. "[455]

From the mid 2000s to the latter half of the year 2020 Musk was a resident of California in the same state where Tesla as well as SpaceX were founded and where their headquarters remains. In 2020, he relocated to Texas after stating that California had grown "self-satisfied" of its financial achievement. In his role as a facilitator the show Saturday Night Live in May 2021, Musk expressed that he is afflicted with Asperger's syndrome.

Chapter 10: Developments And Innovations

Hyperloop

In August of 2013, Musk delivered an idea for a new kind of transport system called"the "Hyperloop," a development that encourages driving between cities with significant populations while drastically reducing time for travel. In a perfect world , completely unaffected to weather conditions and controlled by green power sources, the Hyperloop will propel passengers into units via an array of tubes with low pressure at speeds exceeding 700 miles per hour. Musk discovered it was that Hyperloop could take between 7 to 10 years build and ready to use.

Despite his claim that he outlined the Hyperloop with the claim that it is much more secured than train or plane and would have an expected cost of $6 billion -- roughly 1/10th of the cost of the rail framework that was arranged by the state of California Musk's concept has been met with suspicion. However Musk, the business visionary has tried to convince people to support the idea of improving it.

Following his announcement of a contest to groups to submit their ideas for an Hyperloop prototype, the primary Hyperloop Pod Competition was held at the Space-X office in January 2017. The Speed record at 284mph was established by an German understudy design group in the rivalry No. 3 of every year and a second group setting the record at 287 mph in the year following.

Neuralink and man-made intelligence

Musk has sought to gain an desire to study computer-generated reasoning, and has been elected co-chair on the benefit-free Open AI. The think-tank formed in the last quarter of 2015 with the stated goal of developing computerized insights to benefit humanity.

In 2017, it was revealed that Musk was a major investor in a venture named Neuralink which aims to create devices that can be incorporated into the human brain and help those who struggle with programming. Musk conceived the company's development during a conversation held in July and revealed that the gadgets consist of a small chip that connects via Bluetooth to a mobile phone.

Fast Train

In the latter part of November 2017 following Chicago Mayor Rahm Emanuel made a request to construct and build an express rail line that could transport passengers to and from O'Hare Airport to downtown Chicago within a short time or even less than that, Musk tweeted that he is not averse to his opposition to The Boring Company. Musk said the concept for the Chicago circle was different from his Hyperloop and its typically short course is not requiring the need to attract an air vacuum to eliminate the air-screwing.

In the summer of 2018 Musk stated that he would fund the estimated $1 billion required to build the 17-mile long burrow from the airport up to the city of Chicago. In any event In late 2019, Musk tweeted that

TBC will focus on the completion of the business part of the burrow located in Las Vegas prior to going to other activities, and suggested that the plans for Chicago will remain in a state of in-between for the moment.

Flamethrower

Musk has also apparently discovered an opportunity for business with Flamethrowers from The Boring Co. Following the announcement that they were on sale at $500 each, in late January 2018, Musk claimed that he had sold 10,000 of them in the span of a single day.

Relations with Donald Trump

In December of 2016, Musk was named to the President's Strategy and Policy Forum; in January of the following year Musk joined President his president's Manufacturing Jobs Initiative. In the wake of Trump's political choice, Musk ended up on his side with the newly elected president and his advisers when the president presented plans to pursue massive improvement in the framework.

As he was at possible issues with President Obama's unpopular actions, like the proposed ban on people from Muslim-majority countries, Musk protected his association with the organisation. "My goals," Musk tweeted in the middle of 2017 "are to accelerate the progress of humanity towards economic energy, and also to aid in creating a multi-planet human civilization as the outcome will lead to the creation of a multitude of job opportunities and a truly enthralling future for everyone."

On June 1, after Trump's announcement that he would be withdrawing from the U.S. from the Paris environment agreement, Musk ventured down from his job warnings.

Individual Life

Children and spouses

Musk has been married twice. Musk got married Justine Wilson on the 21st of December, 2000. The couple had six children together. In 2002 the first child of theirs jumped the bucket when he was just 10 weeks old after suffering from sudden infant passing syndrome (SIDS). Musk as well as Wilson had five additional children twins Griffin and Xavier (brought to this world during 2004) and three children Kai, Saxon and Damian (brought to the world in the year 2006).

After a disagreement with Wilson, Musk met entertainer Talulah Riley. They got married in the year 2010. They broke up in 2012 but they were married again in 2013. Their relationship ended in divorce in 2016.

Musk was rumored to be dating singer Amber Heard in 2016 subsequent to ending his divorce with Riley and Heard was able to end her separation of Johnny Depp. Their hectic schedules caused the couple break up in August 2017, but they reconnected in January of 2018 and broke up a month later.

In May of 2018, Musk started dating performer Grimes (conceived Claire Boucher). The following month, Grimes declared that she switched her moniker to "c," the image for speed of light. She claimed to be to be a consolation for Musk. People slammed the women's activist entertainer for a relationship with a billionaire who's organization was portrayed as an "hunter zone" with allegations of unsuitable conduct.

The couple spoke about their affection for each other in a piece published in March 2019 published in Wall Street Journal Magazine, with Grimes declaring "Look I love him. He's great...I believe he's just an amazing goddamn person." Musk in the sense that is important to him, stated to in the Journal, "I love c's crazy fae creative imagination and extremely hard-working attitude."

Grimes gave birth to their child in May 4th, 2020 and Musk saying that the couple named the child "X A-12." Then after the child was identified for the fact that it was discovered that State of California wouldn't acknowledge names with numbers the couple stated that they would change the name of their child to "X A-Xii AE."

Conclusion

Beginning with humble beginnings, a child who was unable to make acquaintances, Elon Musk has risen to become one of the most influential individuals around the globe. A truly gifted genius with a knack to harness human potential his story is one to marvel at. He was presented with a challenge in which he had to construct items to be a part of the world was his dream to live in. So that's what he did.

There was no plan for a space trip when Musk was growing old and neither was there any attractive electronic vehicles. But, Musk dreamed of the travels he read about in his favourite books and thought it would be amazing for human beings to be interplanetary and devastating if they couldn't. At a time when nobody could even imagine investing in private space travel and he started the first space company by himself. In the present that he has more than 40 launches successful and has been the driving force behind the current technological advancements in the space industry.

There's plenty to take away from this tale. Musk wouldn't have been the person he has become without having faith in himself and committing himself to his work. It's true that good things require effort so if you're willing accept that you're up to the opportunity to become an influential , successful individual. Develop yourself, learn to budget and execute your ideas. Nobody else is going to take over your task So you may be wise to believe in yourself and do things your way.

www.ingramcontent.com/pod-product-compliance
Lightning Source LLC
Chambersburg PA
CBHW050402120526
44590CB00015B/1799